simsoc

simulated society

THIRD EDITION

COORDINATOR'S MANUAL
with Complete Materials

WILLIAM A. GAMSON

with the assistance of Russell Stambaugh

THE FREE PRESS
A Division of Macmillan Publishing Co., Inc.
NEW YORK

Collier Macmillan Publishers
LONDON

The Free Press
A Division of Macmillan Publishing Co., Inc.
866 Third Avenue, New York, N.Y. 10022

Collier Macmillan Canada, Ltd.

Library of Congress Catalog Card Number: 77–84284

Printed in the United States of America

printing number

2 3 4 5 6 7 8 9 10

Library of Congress Cataloging in Publication Data

Gamson, William A
    SIMSOC : simulated society : coordinator's
manual with complete materials.

    Bibliography: p.
    1. Social sciences--Simulation methods.
2. Game theory. 3. Social control. I. Title.
II. Title: Simulated Society.
H61.G269 1978   300'.1'84   77–84284
ISBN 0-02-911180-3

Contents

2 BLANK FORMS AND OTHER MATERIALS 71

FIGURES, TABLES, AND SAMPLE FORMS

Blank Forms and Other Materials

Form P: Coordinator's Summary of Operating Information (4 copies)

Form Q: SIMSOC Roster (4 copies)

Form R: Regional Summary Sheet (8 copies)

Form S: BASIN List of Passages

Form T: Key for BASIN Passages

Form U: RETSIN Anagrams and Marketable Words

Form V: Investment Tally Sheet

Form W: National Indicator Calculations (13 copies)

Form X: Basic Group Income Calculations (14 copies)

Form Y: Report to MASMED (16 copies)

Form Z: Special Events (3 copies)

Special Event: Massive Foreign Aid Offer

Special Event: Expeditionary Force

Special Event: Epidemic in the Red Region

Special Event: Earthquake

Special Event: Foreign Threat

Form AA: Coordinator's Questionnaire

Simbuck Sheets—$1

Simbuck Sheets—$5

Simbuck Sheets—$10

Simbuck Sheets—$20

Travel Ticket Sheets

Subsistence Ticket Sheets

Munchie Ticket Sheets—1M

Munchie Ticket Sheets—5M

The "Blank Forms and Other Materials" section (tear-out sheets) starts on p. 71.

Note: Due to their need for special tear-line punching, Forms R, S, and Y appear between Form AA and the $1 Simbuck Sheets.

Acknowledgments

This third edition of SIMSOC represents a substantial improvement over the previous edition. Much of the credit for the improvement rests with a brilliant game designer, Russell Stambaugh, with whom I have collaborated in this revision. Stambaugh was responsible for a number of suggestions that helped to solve fundamental design problems in the previous version of SIMSOC. We discussed all of the specific rule changes in detail. In addition, the explication of the model underlying SIMSOC, a long-overdue task, was performed with his full collaboration.

The second edition, like its predecessor, was widely used. It is a reasonable guess that SIMSOC has now been played by some 5,000 groups or classes, including close to 200,000 participants. Only a tiny fraction of these experiences are known to me. The second edition contained a questionnaire for the coordinator to fill out, describing his or her experiences in running the game and making suggestions for the improvement of the game. This version includes a similar questionnaire. Dozens of people took the trouble to fill out this questionnaire and send it to me, frequently with accompanying letters including suggestions for rule changes. Many of the changes in this new edition reflect these suggestions. So many people were helpful in this regard that it would be unwieldy to acknowledge them individually here. However, I have tried to give credit in footnotes to the text of this *Coordinator's Manual* when a particular suggestion has been adopted.

I am grateful for the very large amount of help I have received in revising the second edition. While this new edition eliminates or reduces many problems of the past, undoubtedly some will remain. I hope users of the game will continue to inform me of their experiences and to suggest changes for incorporation in future editions. Meanwhile, I hope this edition will provide an exciting and valuable learning experience for the coordinator as well as the participants.

WILLIAM A. GAMSON
May, 1977

simsoc

simulated society

1 INSTRUCTIONS

This section may be skipped if you are new to SIMSOC. If you have used the second edition, this summary of rule changes will alert you to the major differences between the previous and present editions. It also includes a discussion of reasons for the changes. Clarifications of rules that do not involve substantive changes are not included here.

There are three types of design problems toward which the changes in this edition are directed:

1. Clarifying the relationship between the private life of an individual citizen and public life in the society.

2. More effectively integrating the pursuit of basic group goals into the ongoing activities of the society.

3. Simplifying and sharpening some of the old rules and adding new ones.

RULE CHANGES CONCERNING PRIVATE LIFE

1. *The problem of lack of personal consumption.* In the previous edition, a luxury living option was suggested (including vacation regions) to give participants a private consumption alternative. It was a weak and ineffective alternative, leaving most individuals with little option other than to use their surplus Simbucks for investment in public programs.

In this new edition, the Permanent Subsistence Certificate has been replaced, and a Munchie Bazaar has been introduced. Instead of purchasing a Permanent Subsistence Certificate, participants may purchase a Luxury Living Endowment, which gives them subsistence plus a supply of Munchie tickets. These Munchie tickets can be exchanged for various snacks at the Munchie Bazaar, run by the coordinator. The number of Munchie tickets available to the society is limited, and the total is affected by the Standard of Living Indicator. If sufficient Munchie tickets are available, individuals may purchase them with their surplus Simbucks if they should so choose.

2. *Narrow range of meaningful individual decisions.* In the previous edition, there was a limited number of choices that individuals had to make, and the effects of many of these choices were quite indirect. For example, an individual supporting a political party might decide to work for that party or even contribute money to it but could easily choose to ignore the political parties altogether. In the current edition, each individual has a number of choices in each session as to which political party to support (if any), whether or not to join EMPIN, whether or not to subscribe to MASMED, and whether or not to declare that he is (or is not) meeting his individual goals. Each of these choices has direct effects on the income of the groups in the society or on the National Indicators. Therefore, the choice that each individual makes is very meaningful to others and is a matter of considerable concern to them, including the choice to ignore the basic

groups. In the previous version, if an individual ignored these groups, it was a matter of little concern.

3. *Quality of work.* In the previous edition of SIMSOC, there was little need for productive labor. The industries solved anagrams, a mildly interesting task which could be performed by the head of the group if he or she wished to do so. In the present edition, there is a great deal more need for labor, and the quality of work can be seen as falling into three categories:

a. BASIN extracts vowels from short passages. This represents unskilled and quite dull labor.

b. RETSIN (which replaces INNOVIN) produces marketable words from anagrams. This represents slightly more interesting and skilled labor.

c. Groups such as POP, SOP, EMPIN, and MASMED have a need to gather individual support cards, a task which can require a substantial amount of travel and persuasive ability.

4. *Individual goals.* In the previous edition, individuals were asked to choose a personal goal and to fill out an Assessment Form which indicated whether they were meeting the goal. There were no real consequences in the game for ignoring individual goals, and the Assessment Form was ineffective in increasing the salience of such goals. In the new edition, individuals have an option during each session of turning in a Goal Declaration Card that specifies whether or not they are meeting their individual goals. These cards affect the National Indicators. Hence, the choice that the individual makes in this regard has important social consequences for others and is of general societal concern.

While private life in SIMSOC has been made more prominent by these changes, the major emphasis of the game is still on problems of public life. The game deliberately departs from isomorphism with a real society in this regard. The amount of labor required to produce the products for the two industries remains deliberately small, and can be performed while still leaving large amounts of discretionary time for virtually everybody. Even individuals performing unskilled labor will not have more than a small percentage of their lives in SIMSOC occupied by such labor.

RULE CHANGES CONCERNING THE BASIC GROUPS

1. *POP and SOP.* In the previous edition, the income of these groups was dependent on the National Indicators but not on their own success in attracting support in the society. In the new edition, they are still affected in the same way by the National Indicators but their base income, instead of being constant, is now dependent on the number of support cards turned in by individuals to the coordinator. Thus, they are forced to garner active support to earn their income.

2. *EMPIN.* The same rule change applies to EMPIN except that it has no competitor in the manner of POP and SOP. EMPIN's income depends on the number of membership cards turned in. Since only memberships of *employed* members of the society are counted toward EMPIN's income, this group now has a

direct stake in keeping unemployment at a minimum, a stake that it did not have in the previous version.

3. *MASMED.* A number of changes in this edition affect MASMED, including several additional perquisites. MASMED's income, however, is dependent on the number of subscriptions turned in to the coordinator in a manner similar to that for POP, SOP, and EMPIN. To receive its income, MASMED must persuade members of the society to subscribe.

4. *BASIN and RETSIN.* RETSIN (Retail Sales Industry) replaces INNO-VIN, from the previous version. The manufacturing process for these two industries is somewhat simpler than in the previous edition, and the work process has been changed in the manner described above under "Quality of Work." The functions performed by these industries are unchanged from the previous edition.

5. *JUDCO.* There are no changes here. While all the other groups now have greater dependency on other members of the society in order to receive their income, JUDCO retains a basic income that is dependent on the National Indicators but not on its performance in the society.

RULE CHANGES FOR SHARPENING AND SIMPLIFICATION

1. *Simforce removal.* The rules governing battles between Simforces have been simplified and made more direct. They were unnecessarily complicated in the previous edition.

2. *Simriot.* A rule allowing members to lower the National Indicators by engaging in a riot has been added. This gives an additional weapon to alienated or discontented members of the society. However, the effects of unemployment on various National Indicators has been reduced somewhat, so that the potency of the strike weapon is correspondingly reduced. Furthermore, the rules provide ways of preventing a riot by the posting of a guard. On balance, the Simriot adds an additional option for making trouble but does not significantly alter the total possible impact of such activities from the previous edition.

In addition to these rule changes, a number of clarifications have been added to the rules to deal with various situations that might arise. These include specific prohibitions against stealing, forgery, counterfeiting, and physical force.

COMPARISON OF THE THIRD AND SECOND EDITIONS

I have had an opportunity to witness two runs of SIMSOC using this new version of the rules. The resemblances between the old version, using a luxury living option, and this version are likely to be quite high. But there are subtle and important differences. The most striking is likely to be the enlarged role of the basic groups, particularly the political parties.

I believe this version is likely to sharpen processes already present in previous ones.

CHANGES IN THE COORDINATOR'S MANUAL

Much of this *Coordinator's Manual* will be familiar to those who have read the earlier edition. Some sections have been expanded and updated. The most significant change includes a lengthy statement of the underlying model of SIM-SOC, with suggestions for using this model in the post-game discussion. (The description of the model is detachable, and coordinators are hereby given permission to reproduce the model for distribution to participants after they have completed the game.)

 Introduction

This manual assumes that you have already reviewed *SIMSOC: Participant's Manual* and have at least a general familiarity with the rules. Here you will be given detailed instructions for each stage in running SIMSOC, forms and materials for doing it as easily as possible, a feeling for what SIMSOC is like in action, and a description of the underlying model.

It is useful to have an assistant or two to aid in the mechanical tasks; some coordinators simply ask for volunteers from the participating group who are willing to assist and observe rather than play. The coordinator's task is more than mechanical, though. There are points where you must make choices that can have important effects on the course of the society. But the major and most important role of the coordinator is to channel the interest generated by the game into the kind of analysis and observation about social phenomena it is intended to stimulate.

THE VALUE OF SIMULATION AS A TEACHING DEVICE

For those who are not familiar with game simulations, the selections by Abt and Raser, reprinted in *SIMSOC: Participant's Manual,* provide a quick introduction. For those who wish to read further in the general area of games and simulations, a brief bibliography is provided in this manual.

Drawing heavily on Campbell (1966), Raser emphasizes the importance of the kind of "knowing" which comes from the recognition of patterns. "We confront a collection of fragments—bits of punctiform data, each of which is uninterpretable—and suddenly we see the entire pattern or context. Common expressions used to describe this experience include 'insight,' 'revelation,' 'seeing how it all fits together,' or 'having it suddenly all make sense.' They all express the recognition that when an entire context or pattern is grasped, each part of the pattern is also more clearly apprehended" (Raser, 1969).

It is not obvious that our traditional teaching methods are effective in achieving such pattern recognition. But there is no need for me to rehearse the familiar litany of criticisms of contemporary education. The criticisms frequently have an underlying theme—education fails when the student is a passive recipient rather than an active participant in the learning process. As Raser (1969) puts it, "The mind should be viewed as an instrument to be honed or tuned, rather than as a bin to be filled. The goal of education should not be to create 'the learned' but to develop the learners. . . . If an environment can be created which will open the mind, stimulate inquiry, arouse curiosity, and provide resources for finding answers, the task will have been accomplished."

Games have great potential as a tool for achieving these educational objectives. Anyone who has used them in teaching can attest to the excitement and interest they generate. This interest must be channeled in an analytic direction by a skillful instructor if it is ultimately going to lead to learning, but the problem of engagement in the process has been virtually solved from the outset. Furthermore, games are particularly useful devices for focusing on "process and system behavior, rather than on the accumulation of detailed bits of knowledge" (Raser, 1969).

Alger (in Guetzkow et al., 1963, pp. 152–54) summarizes the claims of those who have used a variety of game simulations for educational purposes, claims that I would make for SIMSOC as well:

1. It is reported that simulation heightens the interest and motivation of students in several ways. [It is more enjoyable than conventional learning situations, it involves students in ways where their knowledge is relevant to immediate application, and it gives participants a shared intellectual experience with fellow students which frequently carries over out of class and stimulates spontaneous discussion of relevant issues.]

2. It is claimed that simulation offers an opportunity for applying and testing knowledge gained from reading done in connection with the course and from other sources as well.

3. There are numerous ways in which simulation experience is reported to give participants greater understanding of the world as seen and experienced by the decision-maker.

4. Most simulations provide a miniature world that is easier for the participant to comprehend as a whole than are the real institutions themselves.

The last argument is echoed by Coleman (1966), who likens a game to a "caricature of social life." A game may provide for the student of society "that degree of abstraction from life and simplification of life that allows him to understand better certain fundamentals of social organization."

It is extremely difficult to assess these claims, and hard evidence for them remains scanty. Skeptics such as Avedon and Sutton-Smith (1971) are essentially correct when they conclude, after reviewing the evidence for the effectiveness of educational games, that "what is now *known* about games and simulations as instructional media is this: *they are useful devices for getting and holding student interest and attention.*"

The case for stronger claims remains plausible but unproven. The claims for learning are plausible because they make sense in terms of a more general theory of education, and because they are strongly believed by a number of independent users of this teaching technique. But this merely establishes their plausibility, and no more. On the other hand, there is no evidence that students learn less in simulation classes than in conventional classes. Because most of them enjoy such learning experiences more and believe they are educationally valuable, perhaps the burden of proof ought to be on conventional teaching methods.

THE VALUE OF SIMSOC IN PARTICULAR

A contrast between two types of educational games will help to place SIMSOC in context. Some games have a "programmed" quality: the player operates in a situation in which the constraints of the environment are so overwhelming that the player's choice is more apparent than real. If he makes certain choices, he will do poorly; if he makes other choices, he will do well. Such games are interesting when the effects of environmental constraints are subtle and complicated. The educational value consists of developing an understanding of these constraints and how they operate to determine choices. This may be an effective way of teaching the interactions among variables and of giving a player caught up in a process some awareness of the forces operating on him and others in analogous situations.

SIMSOC, however, is a different type of game. The environment it attempts to simulate produces dilemmas or problems for the players rather than forces which determine their behavior. There are many alternative ways of playing it that will, in some sense, "work" perfectly well. The environment is minimally "programmed" to channel the players' behavior in a particular fashion; rather, it attempts to provide the stimulus for sustained exploration and learning. A major design challenge has been that of keeping forces in balance—forces between conflict and common interest, for example—so that no single course of action appears obvious or best for all players.

THE CONTENT

Judging from the list of the courses and settings in which SIMSOC has been used, many people have found it useful in exploring issues that were not central to my own purposes in originally developing it. It has been used in college-level courses in introductory political science, social movements, general education, introductory sociology, police science, social foundations of education, conflict theory, Southeast Asian studies (to focus on the politics of development), human relations, deviance, criminology, communication, interpersonal behavior, group dynamics, community organization, and many others. It has been used with high school students, college students, graduate students, high school teachers, governmental officials, college administrators, prison inmates, nurses, and other adult occupational groups. It has been used in adult workshops in such settings as Esalen and the National Training Laboratory. The city of Plainfield, New Jersey, has hired an experienced SIMSOC coordinator to run regular SIMSOC workshops in human relations for groups of city officials.

Although this varied use pleases me, it also raises questions about just what is the central content of SIMSOC. The game seems to offer at least three different opportunities. A given user may be interested in all three, but typically one receives primary emphasis. The handling of the post-game discussion and written assignments may be quite different depending on which purpose is central.

1. *SIMSOC is used to gain insight into processes of large-scale conflict, protest, social control, and social change.* This is the major purpose that I saw in the game and one that makes it especially relevant to courses that emphasize this content—introductory political science and government, introductory sociology, introductory social psychology, political sociology, community organization, and the like.

The readings in sections 2, 3, and 4 of *SIMSOC: Participant's Manual* by Roberts and Kloss, Alinsky, Warren, Gamson, Dahl, Sykes, Coser, and Boulding deal with this content; most of the suggested study questions in the previous edition focused on these issues. This remains, for me, the central purpose of the game, but I now recognize that there are additional and equally valid uses.

2. *SIMSOC is used to explore interpersonal feelings, communication, trust, and other aspects of face-to-face interaction.* This is the major value of SIMSOC for many people. The game frequently produces highly intense interactions, and participants report surprise at their own behavior and feelings. Thus, it becomes a tool for workshops or courses in human relations, communication, leadership training, and the like. The coordinator functions much as a process-oriented group leader or trainer would in the post-game discussion. The major focus becomes what the participants felt toward each other and toward themselves, why they acted as they did, and other aspects of the content of the immediate interaction.

SIMSOC's focus on large-scale societal processes is really not too relevant for this purpose except insofar as such processes provide the impetus for interpersonal interaction. SIMSOC becomes the content *sui generis* rather than being treated as a model of some outside referent system. Those who use it for this reason typically have a general interest in the simulation technique itself for generating rich interpersonal material. Other simulation games emphasizing quite different content may be explored by such users in addition to SIMSOC.

I have tried to give greater recognition to this use of the game in the reading selections and in the advice in this manual. The readings include selections by Blau, Verba, and Weinstein on "Interpersonal Influence and Leadership" (see *SIMSOC: Participant's Manual*, pp. 97–112), but these tap only one small aspect of interpersonal relations—the one most closely related to the central content of the game. Books such as *The Interpersonal Underworld* by William Schutz (1966) and the literature on encounter groups, on communication processes, and on the dynamics of the classroom (the metagame within which SIMSOC occurs) are equally relevant for this purpose.

3. *SIMSOC is used to allow students to explore the challenge of creating utopia.* One attraction of SIMSOC for many college students is the opportunity it provides them to invent or create a society which they can respect more than their own. SIMSOC is not a society when the game begins. It is merely an aggregate of individuals, some of whom control resources of general social value. Their goals make them interdependent, but they do not yet have collective goals or

mechanisms for making collective decisions. The constraints that the game provides are part of the challenge, but participants are free to create an almost infinite variety of social and political institutions with different mixtures of individualistic and collective values. They import their values to SIMSOC and, working within the sometimes severe obstacles the game provides, they attempt to tailor institutions to fit those values. If there are sharp value conflicts, they may have difficulty in doing this; if they are relatively united on their ideas of what a society should be, they are frequently successful in meeting the challenge of a more cooperative, communal society. The articles in *SIMSOC: Participant's Manual* by Sarason and Slater are included because of their concern with the challenge of creating a set of institutions that serve human needs.

THE APPROPRIATE LEVEL FOR PARTICIPANTS

SIMSOC has been used with three types of participants—high school students, college students, and adult groups. Success with high school students has been mixed. The game is complex and does not lend itself easily to a simplified version. It assumes a considerable amount of sophistication in the handling of forms and in the ability to anticipate and understand consequences of actions that occur. The instructors who have used it successfully at the high school level have run it with relatively sophisticated, college-bound students, not greatly different from the college freshmen these students will soon be.

Although its primary use is with college students at all levels, it has proved to be a quite effective and interesting educational tool with older people. Weekend workshops for adults frequently produce extremely dynamic games in which the participants display great energy, imagination, and ingenuity.

EVALUATION

Do participants actually learn anything from SIMSOC or do they just have fun? The honest answer is that I do not really know, in spite of a few efforts to find out. Some more systematic evaluation attempts have been made both by me and by others. They have generally been inconclusive. These evaluation efforts are typically plagued by two problems—insufficient control of factors other than participation in SIMSOC, and the difficulty of making a reliable assessment of what SIMSOC is intended to teach. When classes using SIMSOC have been compared with classes that did not use it, it has been difficult to prevent self-selectivity from entering the evaluation and difficult to guarantee that the experiences were standardized in all respects other than participation in the game. Since the teaching aim is to produce "insights" or "deeper understanding," it is difficult to develop measures that are reliable enough to reflect what may be very fine-grained differences. Some *behavioral* differences have been found—a lower absentee rate in SIMSOC sections as opposed to straight lecture sections—but these do not establish differences in learning.

Many instructors believe that their students learn. Donald W. Hinrichs[*]

[*] Donald W. Hinrichs is a sociologist at Gettysburg College.

writes, "I can only say that it was the best thing that I have ever used in any course. The students loved it." Lawrence Alschuler[*] writes, "I can say without reservation that SIMSOC is a worthwhile experience, both for the participants and for the instructor. Following up the experience with questions and discussions which integrate the simulation with course materials takes skill and effort, yet can pay off greatly in terms of enlivening textbook concepts and theories."

Participants, for the most part, also think that they learn. It is worth quoting some of their comments on the SIMSOC experience. These selected testimonials prove nothing but they do illustrate the kind or quality of the learning experience participants feel SIMSOC provides:

"It pushed me farther and farther into wanting to know about the things we were doing on a larger scale."

"People actually went as far as to argue heatedly with others they had never even noticed before in the classroom."

"It made many of the items we read about happen in real life."

"The interaction of reading material with actual person-to-person interaction made the course come alive. . . . It reinvigorated my interest in sociology and psychology."

"It was of much more interest than the readings by themselves."

"It got me to notice things both in SIMSOC and outside."

"I became much more aware of the significance of the readings. This was the lab's major asset. It helped me to see 'real-life' representations."

"Seeing some of the theories in action greatly helped me to understand them."

"It made me think about what was actually happening every day in social interaction."

"The idea of a synthetic society and its potentialities fascinated me."

"It stimulated my interest in this area. I took the course to complete social science requirements but would now like to take (and probably shall) more courses in it."

". . . Made me more aware of watching behavior and interaction in an analytic fashion. . . ."

"I found it had a great reinforcing effect upon principles learned through readings."

The reactions of participants are ultimately inconclusive for establishing whether or not they have learned from SIMSOC, but they do have certain virtues. They are easy to obtain, and it is a good guess that, if someone thinks he has learned something, he is more likely to have learned something than if he thinks he did not learn anything. On the assumption that many SIMSOC users will want to gather this kind of information, a list of questions is provided on the following pages.

---

[*] Lawrence Alschuler is a political scientist at the University of Hawaii.

Suggested Questions for Course-Connected Use

1. Participating in SIMSOC helped to make the material in the readings, lectures, and discussion more meaningful to me.

   (*Circle one*)　　　　　　　**Agree**　　　　　　　**Disagree**

2. I spent more time thinking about (sociology, political science, communication, social conflict, etc.) because of participating in SIMSOC.

   (*Circle one*)　　　　　　　**Agree**　　　　　　　**Disagree**

3. I cannot see much connection between SIMSOC and the rest of the material in the course.

   (*Circle one*)　　　　　　　**Agree**　　　　　　　**Disagree**

4. Because of participating in SIMSOC, I spent more time talking to others about the material in this course than I would have without SIMSOC.

   (*Circle one*)　　　　　　　**Agree**　　　　　　　**Disagree**

5. Participating in SIMSOC actually decreased my interest in (sociology, political science, etc.).

   (*Circle one*)　　　　　　　**Agree**　　　　　　　**Disagree**

6. Participating in SIMSOC made me more likely to take other (sociology, political science, etc.) courses.

   (*Circle one*)　　　　　　　**Agree**　　　　　　　**Disagree**

7. I probably would not take another course that used game simulation as a teaching device.

   (*Circle one*)　　　　　　　**Agree**　　　　　　　**Disagree**

8. Participating in SIMSOC has deepened my long-term interest in (sociology, political science, etc.).

   (*Circle one*)　　　　　　　**Agree**　　　　　　　**Disagree**

9. SIMSOC was just another activity required in the course.

   (*Circle one*)　　　　　　　**Agree**　　　　　　　**Disagree**

10. Participating in SIMSOC made me understand better the real-life applications of the processes discussed in class.

    (*Circle one*)　　　　　　　**Agree**　　　　　　　**Disagree**

**Suggested Questions for
Non-Course-Connected Use**

1. How much do you think you learned from participating in SIMSOC, if anything?

    (*Circle one*)   a. **A great deal**   b. **Some**   c. **Little or nothing**

2. How much involvement did you feel in SIMSOC?

    (*Circle one*)   a. **A great deal**   b. **Some**   c. **Little or none**

3. I am glad that I had an opportunity to participate in SIMSOC.

    (*Circle one*)   a. **Strongly agree** b. **Agree**   c. **Disagree**
    d. **Strongly disagree**

4. I feel confused about what SIMSOC was trying to accomplish.

    (*Circle one*)   a. **Strongly agree** b. **Agree**   c. **Disagree**
    d. **Strongly disagree**

5. Here is a list of scales with an adjective at one end and its opposite at the other end. Check the point on the scale for each pair of adjectives which represents how you felt about participating in SIMSOC.

    Example:

Hot		√				Cold
Comfortable						Uncomfortable
Relaxed						Tense
Upset						Calm
In control						Lost
Insecure						Secure
Confident						Unsure
Powerful						Weak
Favorable						Unfavorable
Active						Passive
Apathetic						Energetic
Soft						Hard

6. What did you find most valuable about participating in SIMSOC?

7. What did you find least valuable about participating in SIMSOC?

SIMSOC SIMSOC in Action

THE THEORETICAL MODEL

At the Start

At the beginning of SIMSOC, one could hardly say that a society exists. In fact, the participants face the task of establishing a social order under a very difficult and trying set of conditions. Let's examine these unpromising conditions in some detail.

1. *Extreme inequality among individuals and regions.* A handful of people control the major resources in the society, including the means of subsistence, communication, and production. They not only possess present resources but have continuing control of future resources if they are able to maintain their positions. Others not only lack resources but are in a state of dependence for mere survival. From the very beginning of the society, there are clearly visible haves and have-nots.

The privileged members of society are distributed by region such that there is one region with some abundance, some more or less self-sufficient regions, and one with nothing. The problem of extreme inequality of power and wealth is not merely a matter of individual differences but is overlaid with "geographical" ones. There are have and have-not regions as well as individuals.

2. *Subsistence scarcity.* There is an imminent societal crisis brewing around the lack of sufficient subsistence for the present population. The problem is not merely one of distribution of subsistence, because there simply is not enough to go around without "importing" additional subsistence from the coordinator. Participants are unaware of this imminent crisis, but it looms on the horizon, and they will soon be forced to deal with it in some way.

3. *Communication barriers.* There are powerful communication barriers between regions (with free communication *within* regions). While some individuals possess the means for overcoming some of these barriers, these means are themselves subject to the unequal distribution described above. It is particularly difficult for the deprived region to communicate with other members of the society, since initial communication must come through outside initiative. Thus, the deprived region is isolated as well as deprived. Communication among other regions is difficult as well.

4. *Non-legitimacy of privilege.* All participants know that the wealthy and powerful hold their positions on arbitrary grounds. There is no ideology to justify this privilege, no sense in which anyone could regard the privileged as more deserving than the unprivileged. While it would be too strong to call the privilege "illegitimate," it is clearly non-legitimate. The privileged hold their positions because they are luckier, not more virtuous or deserving in some way.

5. *Poverty of culture.* There is a general lack of shared experience and expectations. Members of the society have no shared history in the society nor clear expectations about what to expect. They bring shared culture and language with them, but its applicability to the current situation is unclear. Normative expectations are ambiguous or nonexistent and are, at best, very fluid and changing.

6. *Lack of government.* There is no organized capacity for making collective decisions or dealing with problems in a collective way. No individual or group exercises any legitimate authority over anyone. No mechanisms exist for making collective decisions or dealing with crises that might arise.

7. *Lack of infrastructure.* There is no organization below the level of the total society. The basic groups do not exist in anything more than name. They consist only of specific individuals who control resources, and who do not function as groups. No loyalties or support structures for these groups exist. In short, there is not yet any structure to support the functions which the groups have been assigned by the rules.

8. *Diversity of personal goals and lack of clarity on how to achieve them.* Individuals are pursuing a range of different personal goals, some of which conflict with each other or with societal and group goals. The goals they are pursuing are independent of any means provided by the society. Thus, there is no integration of personal goals and social position but rather a general state of uncertainty resulting from the lack of culture and social organization. Most people don't have any very clear idea of how to go about trying to fulfill their personal goals.

By any reckoning, this is a formidable set of problems and constraints. The challenge of dealing with them provides the dynamic by which the game runs.

Nature of the Model

The model will describe the central processes in SIMSOC as involving three phases. Each phase presents the members of the society with a characteristic set of issues or dilemmas. The participants can deal with these dilemmas in a number of different ways, but it is possible to classify these ways more abstractly into a very few. The model suggests why some of these ways are more likely to occur than others. How problems are handled in the first phase affects the precise form in which the issues of the next phase are presented to participants. Consequently, the model is rather complex, for we must describe the possible outcomes of each phase and follow their path into subsequent phases.

The First Phase

The first phase is characterized by a number of processes which occur simultaneously and are resolved in a limited number of ways. The most central process is one of mobilization around the problem of the *provision of subsistence* to the participants—that is, overcoming the problem of scarcity. How this general problem gets resolved depends on the handling of the following more specific problems:

1. *Organization of the basic groups.* The easiest and most common path is for basic groups to organize on a regional basis. Because one is in direct, face-to-face communication with one's fellow region members and must travel to communicate with others, it is most convenient to build an organization around this nucleus. This means that political parties initially will take on a regional character, as will industries and other groups.

This process is one in which individual positions of privilege become located institutionally in a home region. Regional organization and organization of the groups are virtually synonymous in this phase. The infrastructure that develops, being regionally based, increases the distrust between the privileged from different regions. If some of the privileged are also minority group members, this mitigates further against the development of a cohesive elite. The privileged are unlikely to develop a general "ruling class" consciousness or to coordinate the pursuit of their interests in this phase.

2. *Organization of the non-deprived regions.* As indicated above, we expect non-group heads in a region to become employees and, in some cases, to take over the leadership of basic groups operating in their region. Because of heavy interlocks between group memberships within a region, each region will become a complex of the basic groups within it. For example, an industry and a political party may come to act as a single unit in many situations.

There may be internal power struggles between the head of a group and its employees, or between group heads in the same region. Some group heads may sponsor a collective leadership while others may maintain various marks of privilege and power that differentiate them from the other members of their region. In general, given the precarious nature of life in SIMSOC during this phase, we expect privileged individuals to attempt to maintain their position of privilege and to share it only reluctantly. Close alliances with one or two trusted lieutenants seem more likely than region-wide decision-making and control of basic groups. The abundant region (Green) is particularly likely to encounter problems of internal cohesiveness, as most members are wealthy enough to be relatively autonomous. Interdependencies among members will not be too apparent at this stage.

3. *Organization of the deprived region.* While the problem facing the better-off regions is that of the institutionalization of privilege, the deprived region (Red) faces the problem of survival, and under extremely difficult circumstances. The members not only lack subsistence but are isolated by the absence of any means of travel. They have no way of initiating communication with other members of the society, and have little or no internal differentiation among themselves.

Under the circumstances, their organization is likely to focus on issues of getting jobs and subsistence for members. During this phase, issues of control over resources are not yet central, although they may be raised.

4. *Regional integration.* At a collective level, the participants are dealing with the problem of regional integration. In a typically unplanned way, they will establish some pattern of communication (or non-communication) with the deprived region and will develop some form of relationship (or lack of relationship). The manner in which the basic groups and regions become established and interact will determine the outcome of Phase One.

Outcomes of Phase One

We will describe the outcomes of Phase One in terms of a number of ideal-typical patterns, and will attempt to state the conditions which give rise to one or another. In any actual SIMSOC, one is likely to find elements from more than one of the archetypes which we distinguish here analytically.

1. *Organized challenge to privilege.* There are various reasons why we expect this to be the most likely response to the problems of Phase One, with the challenge spearheaded by members of the deprived region. Members of the deprived region are typically left alone with ample time to become aware of their common plight. In the absence of internal differences (except for individual goals), their peril and their concern for survival are conducive to solidarity and a sense of common fate. Their high ratio of in-group interaction combined with little or no communication with outsiders reinforces this. Given the lack of infrastructure and government in the better-off regions, it is extremely difficult for members of these regions to organize a rapid response to the problems of the Red Region, even with the best of will. While members of the other regions are still struggling with the tasks of building an infrastructure, the impatience of the Red Region for some societal response grows rapidly.

Since the privileged hold their positions through no special merit, members of the Red Region are likely to develop a hostile attitude toward the privileged rather quickly. If discussions of the plight of the deprived region are taking place, they are typically invisible to the members of the Red Region and take place without their participation. Thus, they tend to be presented with "solutions" to their problem as *faits accomplis*. Far from sparking gratitude on the part of the Red Region, such efforts frequently increase their resentment and anger. Thus, the very efforts of other societal members to deal with the problem of scarcity are likely to stimulate the confrontation.

The political organization of the deprived region is facilitated by the lack of competing tasks in this region. No one has any basic group business to attend to or any other kind of competition for his attention and energy. In contrast, the members of other regions have individual tasks of various sorts—the disposition of travel and subsistence tickets and the organization of the basic groups—that divert their attention from working on the problem of scarcity. Thus, the problem receives secondary priority in the rest of the society during a period in which it is the sole preoccupation of the deprived region.

In sum, the organization of a challenge to privilege by the deprived region rests on the convergence of two sets of factors: those that make it difficult for the other members to respond rapidly and effectively to the problems of scarcity (communication barriers, poverty of culture, lack of government, and lack of infrastructure); and those that make it likely that members of the Red Region will develop strong distrust toward members of other regions (vulnerability and insecurity from lack of resources, isolation, and the non-legitimacy of others' privileges).

There are also factors working against this dominant outcome. Within the deprived region, there are problems arising from the poverty of culture and diversity of personal goals that may handicap organization for collective action. Outside of the Red Region, there are resource holders who have the wherewithal to make a variety of rapid individual responses to the scarcity problem. These efforts may avert a challenge to privilege. We will briefly examine some of these alternative outcomes.

2. *Collapse of the Red Region.* In this solution, members of the Red Region die or make individual arrangements with power holders in other regions and move into the region of their sponsor. The region disappears and survivors are integrated. We would expect this to happen under the following relatively unusual conditions: *a.* members of the deprived region are so different in their goals and common understandings that they are unable to reach agreement on any collective way of dealing with their common plight, and *b.* resource holders in other regions move rapidly to co-opt emergent leadership in the deprived region. This is most likely to happen when some individual members of the deprived region are involved in prior social relationships with members of other regions, who are themselves concerned enough to look after them and act before the consciousness of the deprived region has had an opportunity to gel.

3. *Unstable dependency.* In this outcome, the population of the deprived region becomes, at least temporarily, a ward of the privileged. Members are provided with jobs and subsistence and, if it is needed to maintain their loyalty, a certain amount of luxury. Control of major societal resources, however, lies completely outside the region.

The factors mitigating against this solution are those which make it difficult for members of other regions to act in a coordinated way—communication barriers, poverty of culture, lack of government, lack of infrastructure, and diversity of personal goals. Furthermore, it is difficult to stabilize this outcome more than temporarily because of those factors, described above, that push the deprived region toward an active challenge to privilege.

The most likely form in which such an outcome might emerge in this phase is through the sponsorship of an individual or region which builds an alliance with the deprived region based on an exchange of subsistence and other resources for political support or work. Members of the deprived region may accept such a solution on a temporary basis to insure their survival, but it is likely to be unstable, merely delaying the challenge of the deprived region. A challenge remains a possibility as long as the forces promoting collective action remain intact. The primary forces here are dependency on the privileged, restricted communication, and non-legitimacy of privilege. These variables can be affected in important ways by the participants, however, so stabilization of this dependency during the second phase is a possible outcome.

Phase Two

The central processes in phase two center on what kind of *political, economic, and social organization* the society is to have. Typically, a power struggle ensues concerning these issues, a struggle that can take two forms:

1. *The revolutionary struggle.* This struggle is characterized by the deprived region's demand for some form of public control of resources and the means of

production in the society and for the abolition of private ownership and control. Typically, the demand is resisted by privileged members of the society, who regard it as unnecessary and uncalled-for. The outcome of this struggle depends on a number of factors that will be discussed below, including the presence of the second type of power struggle.

2. *Struggle among elites.* This struggle is characterized by jockeying for power between different groups of the privileged, typically taking the form of regional conflict. In this struggle, region-based groups or powerful individuals vie with each other for the status of top dog. This struggle may be "ideologized" into a dispute about the political, social, or economic organization of the society, or it may be openly over issues of power itself. The need for support in such struggles will produce a tendency for the groups to pursue some appeal for public support—for example, a plan for raising the National Indicators.

One, both, or neither of these struggles may be present in a given SIMSOC, depending in part on the state of the society at the end of Phase One. Using each of the outcomes from Phase One as a starting point, we will indicate the likely progression from the first to the second phase.

1. *From organized challenge to dual power struggle.* The most likely axes of conflict in Phase Two, when there is an organized challenge from the deprived region, are the occurrence of conflict between the revolutionary movement and the privileged, and conflict within the elite. All of those factors mentioned which make it difficult for the privileged members of society to work together are conducive to the development of conflict and distrust within the elite. The problems of unequal distribution of resources, lack of communication, lack of legitimate authority, and matching individual with societal goals are chronic ones which continue to operate during this phase. Meanwhile, the members of the Red Region, having survived the threat of starvation, will begin to expand the scope of their challenge and press it more forcefully.

In some societies, skillful organization among the elites may enable them to present a more-or-less united front to the challenge of the deprived region. In this case, the power struggle will focus on the single axis of conflict between the haves and the have-nots. Simriots, work strikes, and hunger strikes are especially likely to occur under these circumstances. Such actions are also likely to occur during Phase One, before members of the deprived region have had an opportunity to build any alliances with members of other regions.

It is also possible that the organized challenge of the deprived region will dissipate and fall apart through internal squabbling and external threats and inducements to individual members. If this occurs, then the only axis of conflict will be within the elites, with members of the deprived region playing minor roles as supporters of one or another powerful faction.

Finally, it is possible that an organized elite will face a weak and easily dissipated challenge, and a power struggle will not occur at all or will occur only in highly attenuated form. Such a society would move rather quickly to Phase Three. For all of the reasons that make a dual power struggle most likely, this possibility is the least likely alternative.

2. *From collapse of the deprived region to elite struggle.* In the absence of an organized challenge from the deprived region, the most likely process in Phase

Two is a struggle for power among the other region groups. It is possible, but less likely, that the privileged members of the society will overcome the various obstacles to unified, collective action and will agree on some governmental structure with little conflict. Again, such a society would move rapidly to Phase Three.

3. *From unstable dependency to dual power struggle.* Unstable dependency can easily enough lead to an organized challenge from the deprived region during this second phase. Having secured some means of survival and having built some alliances with members of other regions in the process, the members of the deprived region may become emboldened. Concern about survival during Phase One allows members of the deprived region to accept a subordinate position as a temporary expedient. After this point, their concern turns increasingly from obtaining subsistence and other resources to obtaining a say in the control and distribution of resources in the society.

The fact that the organized challenge emerges somewhat later in this scenario has certain consequences. Survival has depended on support from some members of the privileged regions. The opportunity to play off members of the elite against one another is likely to be present in various forms for members of the deprived region. Thus, a dual power struggle is quite likely to emerge in these circumstances.

But there is also a much larger probability that unstable dependency will turn into stable or institutionalized dependency. If no organized challenge has emerged during the first phase, the dependency relationships established during that phase may grow during Phase Two to maintain the deprived region as a protectorate of a more powerful region or group. In this case, no revolutionary struggle will ensue, although an elite struggle is still quite likely, with the members of the deprived region serving as minor actors in the struggle.

Outcomes of Phase Two

Many of these paths converge on the same outcomes, leaving four possible results of the second phase:

1. *Societal collapse.* If the power struggles of the second phase are prolonged and intense, they may lower the National Indicators to such an extent that the society will collapse. This is most likely in the case of a dual struggle, which is likely to take the longest to resolve. Collapse occurs relatively rarely, however, because the intense crisis caused by the falling National Indicators acts as a significant constraint on the power struggle. Warring parties will frequently put aside their differences in the face of a threat to the survival of all.

2. *Socialist society.* While this is not the most likely outcome, it can occur. It involves the abolition of private control of resources in the society and the creation of some form of public control over their distribution. Typically, the public control will be in the hands of some council or government with representatives from the different regions. In such a solution, it is frequently the case that the deprived region becomes the center of such a government. Ironically, the limited ability of members of the deprived region to travel makes it a logical place

for people to meet and negotiate. Leaders from other regions develop the habit of coming to the deprived region to negotiate, and it becomes the site of many important meetings and decisions.

In order for this outcome to occur, the members of the deprived region must win outside support from people in other regions. They can do this by taking advantage of sources of cleavage in these other regions. For example, have-not individuals in other regions may find the program of the deprived region ideologically appealing. Some members of the elite, conscious of the fragile justification for their own claims on privilege, may join forces with the challenge of the deprived region and lend it tangible support. Or members of the elite may even find it expedient to join a movement that has a clear direction and sense of purpose in a society in which there is little competition in this regard. In sum, the success of the challenge from the deprived region depends on a process of gaining active or tacit support from other members of the society.

3. *Stable unrepresentation or institutionalized dependency.* Again, this outcome is not likely, but it can occur under special conditions. It is not likely to occur as long as an organized challenge from the deprived region persists. In the absence of such a challenge and/or the disappearance of the deprived region, it is the most likely result. Members of the Red Region become supporters of one or another faction and receive a flow of rewards, but without any significant sharing of the control of societal resources.

This outcome appears in several concrete forms which we treat here as equivalent. For example, a member of another region moves into the deprived region and runs his group from that region with other members as employees. Or he continues to run the group from outside but guarantees a continued flow of subsistence, travel, and perhaps even Munchie tickets to the population of the region in exchange for continued quiescence and support. It is even possible to turn a basic group over to a member of the deprived region, as long as this person is without any strong regional loyalty and can be trusted to serve the interests of the sponsoring group or individual.

4. *Mixed outcome.* In this outcome, the deprived region becomes recognized as a co-equal member of the pressure system in the society. It develops a regular say in the control of resources. The economic and social organization of the society continues to include substantial degrees of private control of resources. However, this private control is subject to some public regulation by a political body on which members of the deprived region have proportionate representation. An example of such an instrument might be a council, with representatives from each region, that collects and distributes subsistence, making sure all members of society are provided for, and taxes private-resource holders for the support of public programs.

For various reasons, some of them extraneous to the game, this outcome seems the most likely one in Phase Two. Participants bring their own ideological predispositions into the game, and this may be a more comfortable solution for most American participants than some of the alternatives. However, there are forces within the game itself that may promote it as well. It is difficult for a deprived minority to attack and destroy the position of those with great resources. Success depends on an elite that is beset by infighting, is inflexible and uncompromising or otherwise given to social-control errors, or else is attracted by the ideological appeal of the have-nots.

A more likely elite response is to search for a compromise of some sort that allows for the maintenance of a degree of privilege but with greater limits than before. When such compromises are offered, they are likely to create internal divisions within the deprived region on whether to accept. Splits between "reformers" and "revolutionaries" may develop, thereby weakening the capacity of the deprived region for collective action. Feeling less certain of their internal cohesion and external support in other regions, members of the deprived region may find a compromise solution even more attractive. They may never have demanded more than a sharing of power in the first place, so that proposals short of full public control will not be viewed as a compromise at all but as a full-fledged victory.

The likelihood of such a mixed outcome still depends on some organized challenge from the members of the deprived region. If this region disappears or its challenge dissipates, the outcome of Phase Two is likely to be the institutionalized dependency of the region in a society dominated by private-resource holders in the other regions.

Phase Three

The central process of this phase is the *destruction of the hegemony of raising the National Indicators as the collective goal of the society.* The quality of life in SIMSOC emerges as an issue. To raise this issue is to reassert personal goals as important.

At the end of Phase Two, as long as the society has not collapsed, the characteristic drop in the National Indicators during Phase One and much of Phase Two has halted, and the Indicators have begun to climb. The society has developed some mechanism for allocating enough resources to Public Programs to raise the National Indicators, and has handled most of the problems that tend to lower them.

In the process of solving Phase Two problems, a new line of cleavage develops between "leaders" and "masses." The "leaders" may be people who have had privileged positions from the beginning, since there is no doubt that having resources is a help in becoming an influential figure in the society. On the other hand, some may have acquired resources during the course of the game, and others who possess no resources may still have emerged as influential decision-makers. For whatever reason, certain individuals will have emerged as either a formally designated or an unofficial "government."

These leaders continue to be involved in the societal goal of raising the National Indicators; at the same time they are likely to feel that they are fulfilling their own personal goals. Those who are not central to the political structure in the society begin to question the priority of the National Indicators as a goal. They begin to ask whether a rise in the National Indicators means an improvement in the quality of their personal lives in SIMSOC. The problem for them is boredom. The society is running with the efforts of others, and they have nothing interesting left to do except nibble Munchies.

A reflection of this phase is the development of a leisure culture. Participants begin to develop Simreligions, to bring in guitars and other musical instruments, and to develop recreations of various sorts that are independent of the game.

Some new challenges may emerge during Phase Three based on a redefinition of the requirements of leadership. It is no longer sufficient for an individual or a group that wants support to have a plan for dealing with scarcity or raising the National Indicators (which are problems of the past). Now they must have ideas that will enable individuals to meet their personal goals. An increased number of negative Goal Declaration Cards may start to make Public Commitment (PC) a renewed problem.

The emergence of this phase is a signal to end SIMSOC, so it is not meaningful to talk about its outcome. Once a group has developed sufficient social organization, the problems of running the society can be handled with a small amount of energy by a few individuals, and there is little left for anyone to do. At this point, it is time to discuss and analyze the society in a post-game discussion.

The Use of Force

Force can be used during any phase of SIMSOC but it tends to serve different functions at different times. During Phase One, the most frequent use of a Simforce is a device for resource accumulation. In effect, force is used as a form of banditry with the most likely targets being group heads. A Simforce used for such purposes is usually created by one or two individuals rather than by whole regions or larger groups. During the first phase, privileged members of the society may create a Simforce for personal protection from banditry.

During Phase Two, new Simforces may be created or existing ones used for a somewhat different purpose. Where a power struggle is in progress, force may be used as a way of dealing with a political adversary. In practice, this use may not be easy to differentiate from a vehicle for resource accumulation, since group heads are likely to be involved in power struggles and will still be likely targets of force. However, sometimes the targets will have political influence but little or no resources, and the function of force will be clear.

During the late stages of Phase Two or the beginning of Phase Three, force may be used as a collective device to enforce conformity to some collective decision. In effect, a Simforce becomes a law-enforcement agency, used on behalf of a government to collect taxes, punish deviants, and the like. Only when it is used in this fashion is the victim of arrest likely to be given any prior warning of arrest. The emergence of this type of Simforce will typically involve the conversion of an already existing private force into a public one.

Finally, during Phase Three, the function of a force is mischief. Such a force will typically involve one or two individuals who, while they still have some resources, are no longer linked to governing activity in the society. Perhaps they have a Simforce left over from an earlier stage and use it to arrest a societal leader as a way of showing that they are still to be reckoned with. Or perhaps they have managed to accumulate some resources and, feeling somewhat bored in a stagnating society, they choose, in effect, to throw a bomb to stir up some excitement.

With all of these reasons to use force and the relative ease of employing it, it is not surprising that few SIMSOCs will manage to run their course without any resort to force. The model of SIMSOC is schematically presented in Figure 1.

Figure 1. Schematic Representation of SIMSOC Model

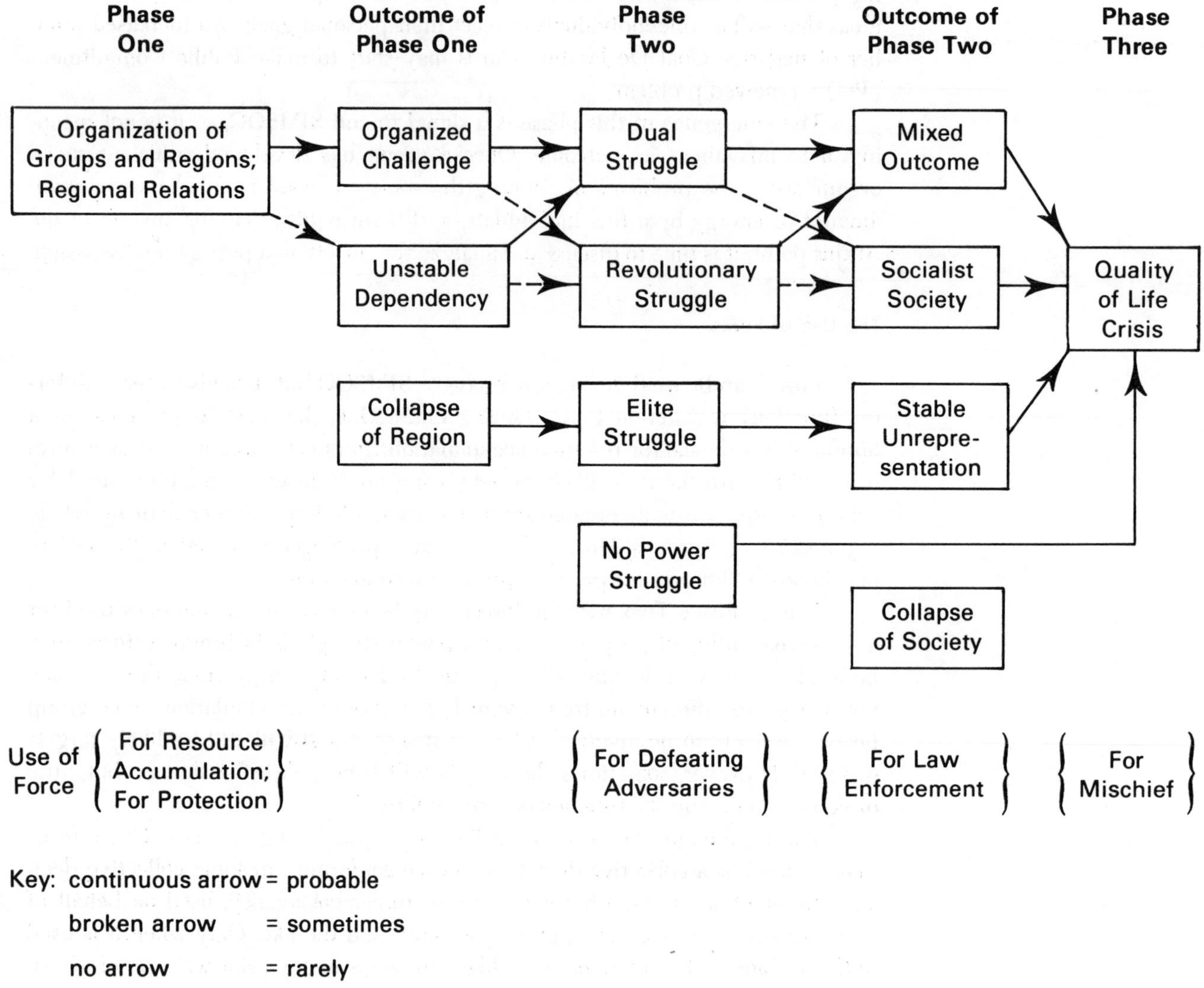

Phase One	Outcome of Phase One	Phase Two	Outcome of Phase Two	Phase Three

Organization of Groups and Regions; Regional Relations → Organized Challenge → Dual Struggle → Mixed Outcome → Quality of Life Crisis

Unstable Dependency → Revolutionary Struggle → Socialist Society

Collapse of Region → Elite Struggle → Stable Unrepresentation

No Power Struggle

Collapse of Society

Use of Force { For Resource Accumulation; For Protection } { For Defeating Adversaries } { For Law Enforcement } { For Mischief }

Key: continuous arrow = probable

broken arrow = sometimes

no arrow = rarely

A HYPOTHETICAL SCENARIO

Although this model gives you a general sense of what to expect, it does not supply the clear sense of SIMSOC in operation that you can get from observing a few sessions. Do not be alarmed, for example, if the first few sessions are somewhat slow. The participants are still learning the rules and getting acclimated during the first session, and many of the dilemmas and challenges in the game take a few sessions to ripen.

As a substitute for direct experience, I will present some imaginary vignettes of characteristic interaction in SIMSOC. Assume that a group of about forty players has completed the discussion of the rules and has asked questions about them. The participants have handed in the Choice Sheet (Form A), indicating their preferences for a basic group and their individual goals. All appropriate preparations have been completed, and participants have gathered in their home regions as the game begins.

ACT ONE, SCENE ONE: *The Red Region. (There are ten people present. Six of them are gathered in a circle; the other four are sitting off to one side looking through the rules. One of the members in the circle opens the discussion.)*

RED ONE: Man, I have no idea what's going on here. (*laughs*)

RED TWO (*after a pause and a few more jokes*): Well, is anyone here the head of one of the groups or anything? (*People in the circle look at their Assignment Sheets again but no one responds. Red Two turns to the others sitting off from the group.*) Are any of you heads of a basic group?
>   (*The coordinator's assistant enters.*)

ASSISTANT: Excuse me. To speed things up, we are providing each region with a list of where the heads of basic groups live and also where the heads of travel and subsistence agencies live. Here's the list. (*He puts* Form R *on the table and prepares to leave.*)

RED ONE: Wait a second. We might have some questions here.

RED TWO (*picks up sheet*): Hey, wow! They've really screwed us. We don't have anything in here.
>   (*Two of the people who were looking through the rules rejoin the group.*)

RED THREE (*to* assistant): We're supposed to get subsistence and jobs, right?

ASSISTANT: Yes, I think so. You die if you don't get subsistence for two times.

RED THREE: Well, how the hell are we supposed to get jobs if there isn't anyone here with any money to hire us?

ASSISTANT: It's a problem. Listen, I've got to drop off these sheets at the other regions. I'll see you later. (*exits*)

RED FOUR: Someone has to go around to some of the other regions to get us some money and jobs and subsistence and things. That's obvious. What society has the most to offer?

RED TWO: Yeah, how are we supposed to travel to other regions when we don't have any travel tickets? Anyway, they're regions, not other societies.

RED FOUR: Doesn't anyone have a travel agency here?

RED ONE: Look, we don't have anything! No travel, no subsistence, no jobs, no money. We're screwed.

RED FIVE: So we're doomed, folks. (*gleefully*) We're all going to die. Let's make the best of it—sing songs, make love, die happy.

RED TWO: Look, someone's bound to come in here. Let's figure out what we want to do so that when they come in we can get what we want. They can't just let us die. It would destroy the society.

RED SIX: Why would it destroy the society?

RED TWO: Some of the whatchamacallits will go down, and if they go down enough, everyone dies.

RED SIX: You mean those National Indicator things?

RED TWO: Right.

> (*A lengthy discussion begins in which the following plan emerges:* Everyone will stick together. No one will take a job unless everyone has a job. They will point out to the other players that if no one is employed, the National Indicators will decline rapidly because of large-scale unemployment.)

SCENE TWO: *The Green Region.* (*The members have learned that they have more than enough subsistence, a travel agency, and the heads of three groups— JUDCO, BASIN, and POP.*)

GREEN ONE (*head of JUDCO*): I need some people to work for me and I also need a subsistence ticket for myself. Anybody here want to do business?

GREEN TWO: We have enough subsistence for everybody, I think.

GREEN THREE (*owner of a subsistence agency*): Three of us have subsistence agencies, so we have more than enough.

GREEN FOUR (*another owner of a subsistence agency*): That is, we have enough for everybody who is willing to pay the price.

GREEN ONE: That's not the right attitude. Trying to make money off people in your own region. You should sell them to outsiders but not to us. We've got to take care of our own.

GREEN FOUR: Well, what do you have to contribute?

GREEN ONE: Money and jobs. Would you like to work for JUDCO? You give me a subsistence ticket and you have a job.

GREEN FOUR: What salary?

GREEN ONE: I only got $45. I'll give you a salary of $10.

> (*A number of other conversations start simultaneously. The heads of POP and BASIN begin hiring people, and the head of BASIN buys a travel ticket for $3 from the owner of a travel agency.*)

GREEN FIVE (*head of BASIN*): I'm going abroad, folks. We need to raise more money to buy passages, so I'm going out to hustle.

GREEN THREE: Hey, you should go to the Red Region. They don't have any jobs in there. Somebody ought to be worrying about that.

GREEN FIVE: That won't do any good. They don't have any money in there. Who lives in Yellow? I don't want to go to my rival's region either. (*Looks at sheet listing group heads by region.*) I'm going to the Blue Region. They have some money in there, and I want to get to them before RETSIN does.

GREEN THREE (*turns to the others*): We have to do something about the Red Region.

GREEN SEVEN: Charity begins at home. Look, I don't have a job, a travel ticket, or nothin'. The society has gotta take care of people like me.

GREEN ONE: You can work for JUDCO, but you have to learn how to take care of yourself. No one is going to take care of you.

> (*The discussion in the Green Region continues, pursuing many different lines and concerns in a scattered fashion. One owner of a subsistence agency leaves with a travel ticket which he has obtained in a trade for a subsistence ticket. He plans to sell his surplus to other regions. The head of POP hires the owner of a travel agency and gives him the job of collecting POP support cards from other regions. No one follows up Green Three's suggestions about the Red Region.*)

SCENE THREE: *The Blue Region.*

BLUE ONE (*head of MASMED*): I would like to announce to one and all the policy of the mass media. We stand for the people. Our editorial policy will be decided by democratic vote, and everyone here who wants to can become a member.

BLUE TWO: What will you pay us?

BLUE ONE: We'll take the income and divide it up equally. I'll take the same share as anyone else.

BLUE THREE (*head of EMPIN*): Right on! The same thing goes for EMPIN. We all share equally.

BLUE FOUR: Hey, great. We're the Blue Commune.

> (*A discussion follows, but the focus remains almost exclusively within the region. There is very little mention or awareness of other regions in the society. Most of the discussion centers around features for the paper. Green Five, the head of BASIN, enters.*)

GREEN FIVE (*with false joviality*): Hello, ladies and gents. I'm your friendly BASIN representative. I'm here to tell you about the wonders of investing in this great company.

> (*The joking tone falls flat and he is greeted by a sullen silence.*)

BLUE ONE: We're not really interested. We don't dig capitalism here, see.

GREEN FIVE (*in a more serious tone*): Yeah, well, that's fine. This doesn't have anything to do with capitalism. We're trying to build the society. We need enough money to buy passages so we can produce our product and the society will grow.

BLUE THREE: Yeah, meanwhile you get rich and buy Munchies while the workers sweat and starve and work their fingers raw counting vowels.

GREEN FIVE (*welcomes the joking tone and relaxes somewhat*): We're a progressive company. The workers will share in all the benefits.

BLUE ONE: Look, buddy. We're not really interested so why don't you try another region?

BLUE TWO: Some of us are interested. He's just speaking for himself.

GREEN FIVE: Great. Why doesn't anyone who is interested come over here to the side, and we don't have to bother the others.

> (*Two members of the group talk with the BASIN head, who decides to hire them. He also asks them to consider the possibility of their moving to the Green Region "if things continue to be so unpleasant in here," and they part amicably.*)

SCENE FOUR: *The Yellow Region. (After a fairly lengthy discussion about the rules and the nature of the game, the members of the Yellow Region begin to focus on the situation of the Red Region.)*

YELLOW ONE: It comes down to this: There are both practical and humanitarian reasons for taking care of the Reds. It will hurt us if they are unemployed, and it's not decent to let them rot in there. They can't even get out to communicate with the rest of us. I think somebody should go in there right away to let them know someone is worrying about their plight.

YELLOW TWO: I agree, but we can't really handle things by ourselves. We need help from the other regions. We could buy some extra subsistence tickets for them but we don't have enough resources to do it all. We should send people out to the other affluent regions to see how much in the way of jobs and subsistence we can line up.

YELLOW THREE: Great. I'll go to the Green Region, and here's a travel ticket for someone to go to Blue.

YELLOW TWO: O.K. I'll go and then I'll come over to Green when I'm through to coordinate.

YELLOW ONE: Why don't you meet in Red? Those poor bastards are sitting in there trapped.

YELLOW TWO: O.K., we'll meet in Red.

SCENE FIVE: *The Blue Region. (Yellow Two enters and waits for conversation to turn to him.)*

BLUE ONE: Can we help you?

YELLOW TWO: I hope so. I'm here about the plight of the Reds.

BLUE TWO: Are you from Red?

YELLOW TWO: No, from Yellow. The Reds have no way of traveling. They don't have a travel agency in there.

BLUE THREE: Well, we're ready to go along. What's the problem?

YELLOW TWO: You know that the Red Region doesn't have any subsistence, jobs, money, or anything, don't you?

BLUE ONE: That's right. We didn't get a chance to talk about this yet.

YELLOW TWO: We think the richer regions have a responsibility to help. Here's a starting plan. Each group hires two of them and pays them at least $5 each for a salary, and we also buy some luxury living endowments so that we have a few extra subsistence tickets to give them.

BLUE ONE: That's cool, but we don't have enough to pay the $5 salary—at least right away. We're only getting $3 ourselves. The rest we have is in a pool for newspaper expenses.

YELLOW TWO: O.K., fine. Let's say $5 starting next session.
> (*Yellow Three enters abruptly.*)

YELLOW THREE: Those bastards!

YELLOW TWO: What happened?

BLUE ONE: Who are you?

YELLOW TWO: She's all right. She's working with me.

YELLOW THREE: The Green Region is a total loss. They won't cooperate, and they can't even get together themselves. Forget them.

YELLOW TWO: It's hard to forget them. They have the extra subsistence tickets.
> (*Assistant enters and announces that the session has 15 minutes to go.*)

BLUE ONE: Let's get over to the Red Region. (*Turns to Blue Three.*) Why don't you represent us?

SCENE SIX: *The Red Region.* (*The Reds have been waiting impatiently for someone to come. They have evolved a definite plan. The coordinator enters.*)

COORDINATOR: There are fifteen minutes left in this session.

RED THREE: Where is everybody? Do they know we exist?

COORDINATOR: They know, but I really can't say anything.

RED THREE: Can you take a message for us?

COORDINATOR: For a fee of $10, or free with the authorization of the head of MASMED.

RED THREE: Can you take a message to the head of MASMED?

COORDINATOR: I'm afraid I can't. You'll have to wait until someone comes.
> (*Blue Three, Yellow Two, and Yellow Three enter.*)

COORDINATOR: Looks like your wait is over.
> (*Great animation and noise in the Red Region. Several people start talking at once.*)

YELLOW TWO: Whoa, whoa, whoa. Listen for a second, will you? Look, we've been worrying about you. The rest of the society except for the Green Region is working to solve your problem.

RED TWO: Why in the hell didn't someone come in here before now? We've been sitting here dying.

YELLOW THREE: We got here as fast as we could.

RED TWO: Someone should have come before now. You should have been including us.

YELLOW TWO: Look, don't you even want to hear what we're planning?

RED FIVE: We have our own plans. Jobs for everybody or jobs for nobody, $10 for every person, and a subsistence ticket for everybody. Take it or leave it.

BLUE ONE: Look, we have jobs for everybody. Will you listen a second instead of shooting off your mouth?

YELLOW TWO: Look, this is your last chance. The session is running out. Do you want to hear our offer, or should we leave?
> (*Several Reds say, "Let's hear it." But Red Five says, "Leave."*)

YELLOW TWO: O.K. Two of you can work for EMPIN, three for MASMED, three for RETSIN, and the rest for SOP. We'll pay everybody $3, and we have four extra subsistence tickets which we'll give to you free.

RED ONE: How the hell could you expect us to pay for them?

RED TWO: That's a lousy offer.

RED FIVE: We don't accept. Our terms are $10 for each person and subsistence tickets for everyone. If you don't meet them, none of us will work, we'll riot, we'll all die, and the society will collapse. Take it or leave it.

YELLOW THREE: You people are incredible. You ought to get together with Green. You don't have any sense of reality. None of us are getting that much ourselves and there just isn't enough subsistence to go around. We are trying our best to help.

RED FIVE: So what do you want us to do, kiss your feet? (*Mockingly, in a fake humble voice.*) "Thank you, Miss. Thank you kindly. Very generous of you. Shuffle, shuffle." (*angrily*) Shove it.

> (*At this point, Green Eight enters.*)

GREEN EIGHT: Excuse me. I have three subsistence tickets for sale. Any buyers?

RED ONE: Yeah. I'll give you a pound of flesh for them.

GREEN EIGHT: I'm serious. If you don't want them, I'll go elsewhere. The price is $10 each.

RED FIVE (*more in pity than anger*): You dumb bastard. How do you expect us to pay you $30 when no one in here even has a job?

> (*The three visitors enter into a negotiation in which they buy three subsistence tickets for $22 and give them to the Reds. Green Eight leaves and the Reds are now somewhat mollified by the last gesture. They agree, with reluctance on the part of some, to accept jobs on the basis that they will get the same share of income as each other employee. In later sessions, the Reds turn their attention to control of the subsistence agencies and insist on their nationalization; a plot to create a Simforce to arrest the head of BASIN is developed by members of the Blue and Yellow Regions. A hostile member of the Blue Region defects, however, and reveals the plot to members of the Green Region, who form their own Simforce first and arrest the heads of MASMED and RETSIN. A considerable battle ensues over several sessions. In the process, the National Indicators sink to a perilously low level but an increasingly effective working relationship develops among members of the Blue, Red, and Yellow Regions. Finally, the head of BASIN is removed from office by his employees and the new BASIN group throws its lot in with the now dominant leadership group. A massive investment program begins to raise the National Indicators and plans for a great national holiday begin to form.*)

This scenario does not, of course, do justice to the variety of activities that are taking place. With interaction going on in four different regions, with sometimes two or three separate conversations in some regions and individuals traveling between regions, it is impossible for any one observer to follow everything. Sometimes (particularly in the early sessions), the society will have no central focus of attention; at other times, a major conflict or crisis will become the center of action. The society sketched in this example illustrates but one of many possible chains of events. Hopefully, it illustrates enough of the characteristic style of interaction to provide a working image of the game in operation.

SIMSOC   How to Run SIMSOC

SPACE

SIMSOC has been run in facilities ranging from one large room to a completely equipped small group lab with one-way mirrors for observation and an intercommunication system for monitoring interaction, making simultaneous announcements to all players, and communicating between regions. Facilities such as student union buildings and gymnasiums have been utilized. Typically, it has been run in less than ideal circumstances, but with a modest amount of ingenuity this has not made a major difference. I will describe the ideal facilities for running SIMSOC and then suggest how to handle various departures from this ideal.

Ideal Facilities

The *perfect* space for SIMSOC would be a well-equipped small-group lab facility laid out as in Figure 2. The salient features of this space are:

1. A *physically separate area for each region* so that players cannot overhear or communicate with players in other regions without traveling.
2. A *central point* through which one must pass in traveling, making it possible for one assistant to collect travel tickets and forms without difficulty.
3. An *opportunity to observe* interaction in the regions without being observed and producing self-consciousness among the players.
4. An *intercommunication system* that enables the coordinator to address people in all regions simultaneously, provides an efficient National Broadcasting System, enables the coordinator to monitor events in any region by the flick of a switch, and allows participants to broadcast proceedings of important meetings to other regions.
5. An *extra area* or lounge that can be used as an uninhabited region for such things as a meeting of regional representatives to create a government or constitution.

Acceptable Facilities

1. *Physical separation of regions.* This feature can be simulated in various ways by asking the players to act as if a physical barrier separated them from other regions. For example, in a large room the four corners may be marked off

Figure 2. Ideal space for running SIMSOC

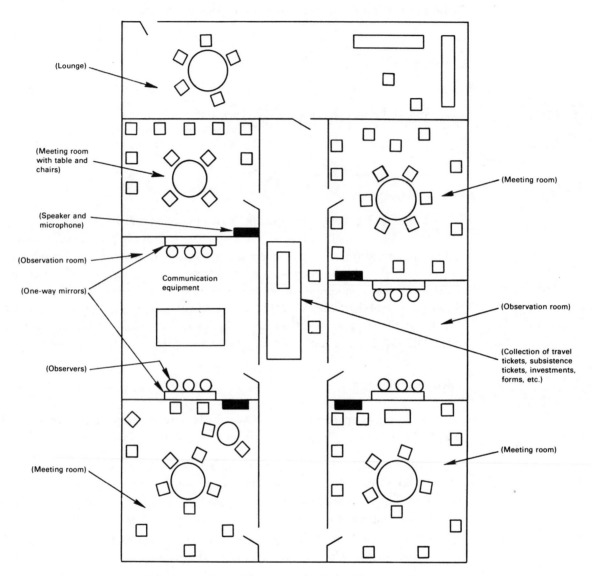

(Lounge)

(Meeting room with table and chairs)

(Speaker and microphone)

(Observation room)

(One-way mirrors)

(Observers)

(Meeting room)

Communication equipment

(Meeting room)

(Observation room)

(Collection of travel tickets, subsistence tickets, investments, forms, etc.)

(Meeting room)

as separate regions, allowing some empty space between regions. Players, of course, are forbidden to call across regions. If possible, curtains should be placed between the regions to afford some semblance of privacy.

Two rooms, although less desirable than four, are better than one. Even a medium-sized room can be divided into zones with a no-man's land in between, and, in such circumstances, one begins to approximate the ideal of four physically separated regions.

2. *A central point in all transportation routes.* This feature is merely a matter of convenience. If it exists, a single assistant can handle the transactions with all four regions with little difficulty. SIMSOC, however, has been run with regions on separate floors of the same building. This required two assistants—one to handle three regions on one floor and a second to handle the region on the other floor.

3. *An opportunity to observe.* A coordinator who moves around from region to region to observe what is happening is likely to find that his entrance affects the interaction. Players are apt to become more self-conscious, and, at the beginning, the coordinator may become a target for questions that he should not answer. This problem is ideally handled by a facility with one-way mirrors and microphones in each region, which enable the coordinator to monitor the interaction. There is no reason to use deception in this regard. One coordinator who has used SIMSOC in such a facility has made a point of giving the players a complete tour of their environment before the game begins and of telling them that they will be observed. Once the game begins, this fact has low salience for the players, and the coordinator can follow what is happening without interrupting it by his presence.

Failing such a facility, the coordinator will do best by remaining fairly stationary, observing a single region for a half-hour or more before moving on. Once the game has some momentum and he has made it clear that he will not be drawn in as a quasi-participant, he may move somewhat more freely. If he has assistants stationed in other regions, he may ask them to describe to him the highlights of what happened in their regions, and thus gain a sense of the society as a whole.

4. *An intercommunication system.* I have rarely had this facility in running SIMSOC, but I have found it possible to approximate it without difficulty. In its absence, messages are simply written out and presented to the nearest assistant for delivery (along with the fee, if required). The assistant then travels from region to region reading the message to the assembled population. Once again, the departure from the ideal physical facility can be compensated for by increased personnel in running the game.

The other advantage of an intercommunication system—the opportunity to monitor what is happening with great flexibility—cannot be had with this device. But the handling of this limitation is discussed above as part of the problem of observation.

5. *An extra area.* It is useful but not essential to have a spare room. In the absence of this extra space, one can simply cordon off a corner of an existing region for the same purpose and treat it as physically separate from the region that inhabits the same room.

TIME

There are three aspects of the time problem: the length of a single session, the total number of sessions, and the proximity of sessions. For each of these issues there is more than one answer, but some ways are more desirable than others.

1. *Length of Single Session.* It is possible but not desirable to run SIMSOC during a standard 50-minute class period. A longer block of time for a session seems preferable, however. Many coordinators like to use their discretion about when to end the session, always after giving at least a 15-minute warning. If participants are especially rushed, the coordinator may let a session run a little long rather than allowing an overly severe time constraint to abort promising developments. If a session seems to drag, he may shorten it somewhat. Some time con-

straint is both realistic and helpful in encouraging discussion to reach a point of decision, but it is useful to have flexibility. The ideal length for a session is probably closer to 75 minutes than to 50. It usually takes participants some time to get warmed up and into the game; if they are too rushed, they will have no opportunity for exploration and innovation and will feel pressure to adopt the first plausible suggestion offered for dealing with a problem that arises.

For classes that are inexorably tied to the 50-minute time period, coordinators might consider using the period as a half-session. This is by no means a semantic distinction. Many things in SIMSOC happen from the end of one session to the beginning of the next. Investments in Public Programs go into effect, payments to the basic groups occur, the effects of subsistence failure go into effect, the National Indicators move, and so forth. To say that a class has ended in the middle of a session removes any real time-pressure for that class period. Of course, the pressure will return in the next class, when the second half of the session takes place, but at least there will have been an opportunity for exploration and innovation before the push for decision and action occurs.

2. *Number of Sessions.* Essentially, one can keep SIMSOC running as long as it seems to be generating rich material for subsequent analysis. It is likely, however, to reach a point of diminishing returns by eight or ten sessions and usually before that. A series of six sessions seems closer to the average for a fruitful run, but a sensible stopping point is up to the coordinator.

In making this decision, the following observations may be helpful:

a. As the model suggests, it takes several sessions for the central processes in SIMSOC to mature. Unless it is allowed to run at least four or five sessions, a good many interesting developments are likely to be aborted. Dramatic events seem to occur most frequently between the third and sixth sessions.

b. It is undesirable to end the game before the struggles of Phase Two have come to some climax. It is also frustrating to the participants to end the game while some long-developing clash is on the verge of resolution. The development of Phase Three issues should be the primary signal that the game has been played long enough for participants to have garnered its primary value.

c. It is not a good idea to announce when the last session will occur. SIMSOC is not an attempt to simulate how people act when the world is about to end. To avoid end-game effects, I have always held the post-game discussion on an occasion in which players assembled for what they expected to be another session of SIMSOC. This is quite possible to do, even if players know it is the last scheduled class. For this reason, I do not say much in the beginning about plans to hold a discussion. Thus, there is no reason for players to doubt that the last scheduled class will be devoted to another session of the game. When players ask which session is the final one, it is best to tell them simply that it will not be announced. It is also possible and perhaps desirable to end the game and begin discussion in the middle of a session rather than wait until the session is completed.

The typical experience, then, is for players to expect at least one more session than actually occurs. The typical reaction to the violation of this expectation is surprise and, for some people, disappointment. If involvement is still high at this point, these reactions may require some skill from the coordinator. Some players are reluctant to trade the role of citizen of SIMSOC for that of social analyst, but, by calling attention to obvious bias in social analysis, the coordinator can usually help most players to recover some detachment, and the high motivation and interest can then be more easily channeled into analysis and observation.

3. *The Proximity of Sessions.* SIMSOC has been run with one session a week, with several sessions per week, and packed into a two-day period which excluded virtually all other activities for the players. It seems to be more effective the shorter the time lapse between sessions.

Participants must work their way into the roles the society provides. At first they are self-conscious, but, as they continue, their role distance diminishes and they use the vocabulary of the game in a natural way. The reality of the society, however, disappears over a few days' interruption by other activity, and they must readjust to the simulated world each time. This means a loss of momentum and a good deal of spinning of wheels at the beginning of a new session.

This process of reintegration is hastened if several sessions are held during the week—for example, on Monday, Wednesday, and Friday. Recently, I have worked quite satisfactorily in a three-hour period on Tuesday and Thursday, running two sessions during each time block. The layoff between Thursday and the following Tuesday involves some loss of momentum, but this happens only once over a two-week period.

The ideal situation, however, is two all-day sessions on consecutive days, usually on a weekend. A typical two-day session starts with a review of the rules from 9:00 to 10:15 on Saturday morning, followed by an initial session from 10:30 until noon. Lunch is served in the regions while the results of the first session are being tabulated. Three more sessions, averaging an hour and 15 minutes each, are run from about 12:45 to 5:15, with 20-minute breaks between sessions.

Two more sessions follow on Sunday morning, and, when players reconvene after lunch, the coordinator announces that the game is over and then leads a two-hour post-game discussion, ending about 3:00 in the afternoon. Involvement is usually quite high under such a program of continual play, and the participants are likely to be rather exhausted at the end if the game has been an interesting one.

In sum, SIMSOC can be run in weekly sessions but, where there are opportunities for putting sessions closer together, it seems highly desirable to use them.

THE USE OF ASSISTANTS

The coordinator's job in running SIMSOC can be made much easier by the use of one or two assistants. These are often drawn from participants. Assisting and observing are probably as valuable learning experiences as playing is. Some people clearly prefer this role, and it should not be difficult to get one or two volunteers from the group of participants.

One of the assistants should be assigned the task of collecting travel tickets. It is best to do this at the beginning of trips. Another should be given the job of collecting subsistence tickets, group support cards, and investments in Public Programs. They can share the task of delivering messages for the National Broadcasting System. The assistants can also help in sorting Choice Sheets (Form A) at the beginning, and between sessions can help prepare the proper amount of Simbucks, travel tickets, subsistence tickets, and Munchie tickets. With two assistants, one can be trained to calculate basic group income (using Form X) while the other calculates the National Indicators (using Form W).

With these matters taken care of by assistants, the coordinator will be quite free to observe during sessions and to deal with any complications and problems that may arise. The coordinator will be kept busy at certain times recording information on the Roster—especially at or near the end of a session, when he may receive several forms at the same time. Assistants who are briefed before the game begins can help to relieve these busy periods.

SUGGESTIONS FOR A SHORT VERSION

It is possible to run SIMSOC in a single all-day period, although this is less than ideal. To do this successfully, you should:

1. *Keep the discussion of the rules quite short.* Participants should read the summary and have a few minutes for questions. Most of the learning of rules will take place during the play of the game itself, with participants reading their *Participant's Manual* as the need arises. The coordinator should be readily available as a resource person on the rules, helping the participants to locate sections of the manual that are relevant to their needs.

2. *Make the sessions shorter*—about 45 minutes to an hour. It is better to get through four or five short sessions than to go through only two or three more leisurely ones.

3. *Do not shortchange the post-game discussion.* Make sure you leave at least an hour and a half. Most of the learning probably occurs in the post-game discussion, and a great deal will be lost if participants play the game without adequately reflecting on and discussing their experience.

4. *Consider the possibility of stretching the time available* by including the previous evening. A game that includes Friday evening and all day on Saturday can be quite satisfactory. A discussion of the rules and the first session can be held in the evening, and four slightly shortened sessions and a post-game discussion can be held the following day. This approximates the ideal situation except for a bit too much time pressure.

NUMBER OF PARTICIPANTS

The instructions provide for a range of 15 to 60 participants. The ideal number is around 40. Too large a group will become unwieldy, and may leave some players with insufficient tasks and roles to involve them in the game. If you have more than 60 participants, it is better to divide them into two or more groups and run them as separate games. For example, 200 students could be divided into four SIMSOCs of 50 each or five SIMSOCs of 40 each.

If the number of participants is too small, another kind of problem arises.

When there are only five or six people in a region and one or two travel on a mission, there is a tendency for the others to cease activity while they wait for the return of the travelers. When the number in a region approaches ten, independent regional activity is likely to continue in the absence of one or two members. This problem can be mitigated in smaller groups by making the minor adjustments suggested in the section "Running an Undersized SIMSOC" (p. 52). Specifically, the instructions call for elimination of one of the four regions with groups of 15 to 23 participants. In other respects, the game remains essentially the same.

ON USING FORMS

The forms are designed to make the running of SIMSOC as convenient as possible. Whenever you don't need to have a form filled out, you can dispense with it. For example, if someone wishes to move and you are unhurried and can record this change immediately on the Roster, you don't need to make the person moving fill out a form.

Some forms are for the convenience of participants. The main ones in this category are those covering Luxury Living Endowments and Private Transportation Certificates (Form D and E). These forms are essentially receipts that the participants can show in the event that their claim is challenged by the coordinator or an assistant. Once you have recorded the information on the Roster, you have no further need for these forms.

The forms enable you to keep track of information that might overload your capacity to record it at any given moment or might not be needed until the end of the session. A good example is the Withdrawal of Assets form (Form I), which enables you to keep track of how much BASIN and RETSIN heads have withdrawn during the session.

Forms are also important in cases where disputes might arise. This is especially true for forms involving conflict events such as arrests, the creation and removal of Simforces, riots, and the removal of group heads. In such instances, forms protect the coordinator from being drawn into internal disputes in the society. Where ambiguities do arise, the forms can be turned over to JUDCO for determination of whether they meet the rules of the game.

I have found the following general procedures helpful in using the various forms provided in this *Coordinator's Manual:*

1. *Record information in pencil, not ink.* This is especially true of the Roster, where the problem is not simply one of correcting a mistake but one of keeping up with changes in employment, region, and the like.

2. *Go to the regions to collect routine materials* such as subsistence tickets, job schedules, and the like. This will avoid congestion at the bank. Participants should come to the coordinator's table mainly to exercise various non-routine options (initiating a Simforce, filing a minority-group action form, etc.), to purchase or convert Munchie tickets, and to drop off travel tickets.

3. In doing all numerical calculations, always *round to the nearest integer.* None of the processes in the game depends on such small differences, and fractions can safely be ignored.

SIMSOC Explaining the Rules

Although each participant should have his or her own *Participant's Manual* before SIMSOC begins, much of the learning of rules takes place while the game is in progress. There is too much information to assimilate prior to play, and, as long as they understand certain basic things, participants can learn what they need to know as they play. Ideally, they will have had an opportunity to read over the whole set of rules before gathering to discuss them, but if they have not been able to do this you should begin by having them carefully read the "Summary of Rules" (pp. 4–7) in *SIMSOC: Participant's Manual*.

Then go through the *Participant's Manual* with them, asking for questions on each section. They will need to read the descriptions of the basic groups and individual goals to fill out their Choice Sheet (Form A). A short Self-Test on SIMSOC Rules (Form O) is provided in the manual. If participants are able to fill it out correctly, this demonstrates that they have sufficient understanding of the basic rules to begin play, although far from enough to cope with the problems they will face collectively.

Certain general points should be stressed in fielding questions on the rules:

1. There is no single right or best way to play the game. There are many alternative ways. Participants should try to keep an open mind about the possibilities available to them; they should not assume that they cannot do something just because no one has thought of doing it before.

2. They should not waste their time trying to figure out what the coordinator expects them to do. What the coordinator expects is that they will use their imaginations and come up with ingenious ways of achieving their objectives, ways that may not have occurred to the coordinator.

3. Try to avoid discussions of how to play the game as opposed to the meaning of the rules. Emphasize that such discussion should properly be held when the game begins. Questions on how to play should, as much as possible, be evaded rather than answered.

4. Remind participants of the distinction between two kinds of rules—man-made rules and natural constraints of the environment. They should accept the rules of the Manual as if they were natural constraints, equivalent to physical ones rather than agreements among men. For example, the requirement that one must have a travel ticket or Private Transportation Certificate to travel should be taken as an analogue of the cost of transportation over long distances in a real society. To travel without a travel ticket is like breaking a physical law (such as the law of gravity)—not like violating the Constitution or an Act of Congress. The success of the game requires that the players observe the constraints in the Manual. But they are not required to observe agreements which may be made by the members of the society. It is up to their fellow citizens to see that such agreements are honored, and the request to obey rules is not intended to cover such man-made agreements or norms.

5. Emphasize the coordinator's role as observer once the game begins. In keeping with this role, avoid as much as possible interpreting the rules where they seem genuinely ambiguous. For example, if you suggest an interpretation and someone points out an ambiguity or an alternative interpretation, do not try to resolve it. Rather, point out the role of JUDCO in clarifying such rules. It is quite proper to suggest that there may be more than one interpretation and that they will have to resolve the issue in the course of the game. This advice applies, of course, only to rules which really are ambiguous. Farfetched and untenable interpretations that are clearly not intended by the rules should be rejected. This means that the coordinator will have to exercise some judgment about when an interpretation of the rules is tenable, but he does not have to choose which of a set of tenable interpretations is the proper one.

I have tried to anticipate some of the most frequent questions on the rules by providing a question-and-answer section in *SIMSOC: Participant's Manual* (p. 30). It may be some comfort to realize that even if you make a mistake in presenting the rules, the game has enough error-tolerance to absorb a certain amount of coordinator error. The additional questions and answers included here illustrate the recommended style of answering. *C* is used for coordinator and *P* for participant:

C: Any questions on the introduction?

P: Well, I'm a little confused about the whole thing. I don't know how to put it in a question exactly. It's more just what the game is trying to get at.

C: You don't need to worry too much about being confused at this point. Things should become clearer as you actually play. If you still feel confused after the game is over, then that's something to worry about, but not at this stage.

P: Why can't there be more than 50% of the players in any one region? What is that supposed to represent?

C: I guess it is a way of reflecting the fact that people in a real society can't all get together for face-to-face communication.

P: Can we meet out in the hall without having it count as travel?

C: No, the hallways and corridors are passages, not regions. You're not allowed to congregate in the halls, with or without travel tickets.

P: Do you have to have a travel ticket to go to the toilet?

C: I don't think that's covered in the rules, so JUDCO would have to decide whether the bathroom is considered part of every region or a region in its own right. I'll make an interim rule that you don't need a travel ticket to go to the bathroom, which JUDCO can overrule later if it wants to.

P: Can we send letters from one region to another?

C: Sure, if you can figure out some way of getting them delivered.

P: What do you mean?

C: I mean that we only provide those services that are specifically mentioned in the manual. Beyond that, you have to provide your own services.

P: You mean we could set up our own postal system?

C: That's right, but your postman would need a travel ticket when he traveled to deliver the mail just like anyone else.

P: Can you work for more than one group?

C: I think so. Doesn't it say that somewhere in the rules? (*Looks up appropriate place in the* Participant's Manual.) Yes, here it is on page 11: "You can work for more than one group."

P: Suppose you tell us one thing and the rules turn out to say something else. Which should we pay attention to?

C: You should pay attention to what the rules say unless I specifically tell you that we are *deliberately* changing one of them. I don't have the rules memorized, you know, and I might forget something and answer a question incorrectly. If I accidentally mislead you, I'm sorry, but you have your copy of the rules to consult when you're in doubt.

P: Doesn't the head of a group have an awful lot of arbitrary power?

C: Perhaps. Is something unclear?

P: No, I guess not, except that in real life the head of an organization doesn't usually have that kind of power.

C: Well, no simulation tries to be an exact replica of the world. The main question is whether some of the ongoing processes in SIMSOC resemble important processes in the real world. That question is one you will have to defer judgment on until after you play.

P: Suppose the heads of groups hire only one other person. Does that mean they couldn't be removed from office?

C: They could be removed by arrest, for failing to provide subsistence, for being absent, or by having a minority-group action form filed if appropriate. However, the one employee couldn't do it. Remember that JUDCO must hire at least two others but the other group heads don't have to.

P: Well, doesn't that mean that most of them will hire only one other person?

C: Wait and see what happens. I can't answer that.

P: Suppose someone lent money to BASIN to buy passages and they didn't pay him back when they promised to. What could he do?

C: Lots of things. What would you do in real life?

P: Sue.

C: Well, there are things that you can do in the game. I don't want to single out anything because that would be doing your job for you.

P: What do POP and SOP stand for?

C: Whatever they choose to stand for in the course of the game.

P: Can a person come back to life by getting a subsistence ticket after he is dead?

C: No, dead is dead.

P: Suppose someone is really sick and can't get to a session. Why should the society be penalized for that by having the National Indicators go down?

C: Just as in a real society, you've got to expect a certain number of legitimate absences. This is something you can try to anticipate and deal with if you think it is important. But the rules say that the absence counts "regardless of reason."

P: If a head of a group is removed for being absent is he entitled to demand his job back when he returns?

C: What do you think?

P: I don't know. It sounds as if it would be up to the new head and the other members of the group.

C: Sure. As far as the rules are concerned, the old head has no claim. However, you may agree to let him back—that's up to you. All we know officially is that there is a new head who gets the income for the group until someone files another Job Schedule changing that.

P: As I understand these rules, you don't need to have a job as long as you have subsistence. If I bought a Luxury Living Endowment, I wouldn't need to work, right?

C: You're right that you wouldn't die. However, the National Indicators would be lowered by your unemployment.

P: That means that it's in the society's interest to make sure that everyone has a job, doesn't it?

C: You should realize by now that I can't tell you what's "in the society's interest." That's something you will have to decide when the game starts. You will frequently find that some of these issues are more complicated than they appear on the surface.

P: Why do unemployment and absenteeism lower Standard of Living?
C: I guess on the assumption that they involve some loss of production for the society.

P: What do you do if you die?

C: Observe, help to run the game, and not participate in any way in playing it. You will not be allowed to give advice or make suggestions or attempt in any way to influence the course of the game. You can learn a lot from observing as well as playing, so dying isn't so dreadful from the standpoint of learning something.

P: Do you have to reveal your personal goals?

C: That's strictly up to you.

P: What do these Public Programs involve? I mean, what happens when we invest in Research and Conservation, for example?

C: There are these four National Indicators that affect how much income the basic groups have. So if you are concerned about this, you need to worry about how the National Indicators are doing. The main way to raise these indicators directly is by putting Simbucks into these two Public Programs.

P: But what happens then? Who carries out these programs? Do you, and what do they involve?

C: There isn't anything actually carried out in a concrete way. It's more like putting money in the bank for these programs. More money in Welfare Services stands for the idea of certain activities which raise some of the National Indicators such as Public Commitment. But these welfare activities are just there in the abstract. You might start your own concrete welfare plan for the people actually here playing the game, but that would be something different from investing money in this Public Program.

P: You said investing in these Public Programs is the main way of raising the National Indicators. Is there any other way?

C: Well, the Public Programs are how you do it directly. But you can also try to prevent things from happening that will lower the National Indicators. And you can do a lot of things that will affect the amount invested in Public Programs. These are actions which indirectly affect the National Indicators, and they may be quite important.

P: If there isn't any government or elections, what's the point of having political parties?

C: The political parties in SIMSOC are organizations of people with a program for the society. If there were elections, the parties might be relevant, but there is nothing about the basic idea of a political party that implies elections or any particular form of government. The political parties we have in the United States are not the only conceivable kind.

P: How do you go about creating a government?

C: That's your problem, and we shouldn't discuss it here.

P: Can a society exist without a government?

C: That would be a good question to write a paper on after you have finished playing. In fact, that's a good example of the kind of question this game is designed to stimulate. Let's save the discussion of it until after the game is over, but it's valuable to keep it in mind as you play.

P: If someone wouldn't do what I wanted, couldn't I just start a Simforce and make him do it?

C: I can't answer that really because making him do something is not part of the rules. The rules say that you could restrict his travel, keep him from holding any official position, and have all his resources confiscated including his Simbucks, tickets, and so forth. They don't say that he has to do anything you want him to do.

P: Why would someone want to start a Simforce except for confiscating someone else's resources?

C: I can't answer that. Let's just say that there are many other possible reasons besides that one.

P: It seems to me that this Simforce rule will mean that the society will turn out to be a police state.

C: Maybe and maybe not. But there's no point in our speculating about it. We'll just have to wait and see what happens. The important thing now is for you to understand what the rules for the Simforce allow you to do, not to anticipate what will happen when you start playing.

P: What kind of special events are there? Does this mean things like wars or earthquakes or things like that?

C: That's the general idea. From time to time, things happen from outside a society that affect it. They happen unexpectedly, and some of you or all of you may have to deal with these things in some way. You are just being warned in advance that such things might happen in this society, but there is not much you can do with this information because you don't know the details. You'll have to deal with such things when they occur.

P: Looking at Table 3 on the National Indicators, it seems that nothing lowers FES. So why would one need to bother investing in the Research and Conservation program?

C: Look again. FES is lowered by two units for each passage bought by BASIN. Furthermore, all the National Indicators automatically decline by 10% between sessions, so FES would go down even without BASIN.

P: Are you going to grade how well we are doing?

C: No. Nothing that has to do with your performance in the game will be graded. Any grade you receive in connection with SIMSOC will be based on your written work and your analysis as a student of the society, not on your skill as a citizen of the society.

SIMSOC Getting Ready to Play

You will find that in preparing to play SIMSOC the biggest expenditure of time is not in the shuffling of paper but in learning and understanding the rules and procedures.

Let me assume that you have read through the rules once for general understanding and have read the selections by Abt and Raser in *SIMSOC: Participant's Manual;* you should continue reading all the way through the *Coordinator's Manual* once. Then return to the rules in the *Participant's Manual* for a more careful reading. The more you have the rules at your fingertips, the more effectively you will be able to deal with situations that arise once the game begins. You can always look things up, of course, but sometimes actions move rapidly, and it is helpful to know just what to look for and where to find it quickly. When you feel that you are adequately familiar with the rules, you are ready to return to this section and begin setting things up to play.

( ) 1. Gathering materials. One needs the following materials to play the game, and it is best to gather them ahead of time.

    ( ) **a.** As many envelopes as there are likely to be participants, plus a few extras in case some people lose their envelopes during the game. If you already know the names of the participants, write one name on each envelope.

    ( ) **b.** Simbucks, travel tickets, subsistence tickets, and Munchie tickets. These are all provided in the back of this manual but should be torn out and sorted into appropriate piles.

    ( ) **c.** Name tags. You should have enough for each player, and a few extras in case participants lose them. It is quite important for participants to wear their name tags. This will ease your job in recording information on the SIMSOC Roster (Form Q) and will facilitate interaction among the participants.

( ) **d.** Coordinator's Summary of Operating Information (Form P). You may find it useful to detach this form from the manual and attach it to a cardboard backing. It will enable you to avoid looking through the *Participant's Manual* to locate the information you need during the course of the game. Extra copies are included so that a copy can be prepared for assistants as well.

( ) **e.** Colored paper to mark region. The easiest way to identify regions is to tape a sheet of paper of the appropriate color over the door. If you don't have red, blue, green, and yellow paper, you can do well enough by writing the name of the region on an ordinary sheet of paper and posting it at a prominent spot at the entrance.

( ) **f.** A hand calculator. You will find it useful to have a calculator with you during the game, although it is not necessary. Most of the calculations are very simple ones, but one or two involve a slightly longer operation.

( ) **g.** Ditto stencils (or typing and carbon paper) plus a typewriter. MASMED must be given materials to produce a news magazine. A typewriter is desirable for this purpose although not absolutely essential. If a ditto machine is available, you may wish to provide MASMED with ditto stencils. If not, one can make do with carbon paper sufficient to produce at least four copies (one for each region). If a photocopier is handy, MASMED can produce a single copy which can then be quickly photocopied for distribution.

Artifacts for Frequent Users

If you run SIMSOC fairly frequently, you may wish to prepare materials that will increase the convenience of running the game. The following are some ideas to consider:

1. Before separating Simbucks and travel, Munchie, and subsistence tickets, glue the sheets onto a *colored paper backing*. Use a different color for each denomination of Simbuck and Munchie ticket, and for travel and subsistence tickets. After they have been securely glued to the backing, use a scissors to separate them into units. One then has the double advantage of heavier, more easily handled materials and different colors to facilitate sorting.

2. Red, yellow, blue, and green *felt-tip pens* are useful for marking the Roster assignments of participants to different regions, enabling one to locate the members of any given region more easily. (But beware the possibility that people may move to different regions, which may spoil the convenience of this system.) Even if you don't use colored pens to mark the region on the roster, you will find it useful for marking envelopes. Since it is frequently necessary to sort envelopes by region, color-coded envelopes enable one to do this more rapidly.

3. A *clipboard* with attached envelopes is useful. You might try taping five small collection envelopes onto the back of the clipboard. They can be used to deposit political party support cards, EMPIN membership cards, MASMED sub-

scriptions, individual declarations, and subsistence tickets. The front of the clip board can hold the Coordinator's Summary Sheet (Form P), the Investment Tally (Form V), and, on top, the Roster (Form Q). It's useful to have this clip-board when traveling to a region. It contains a place to put tickets that may be handed to you in already sorted piles and to store forms; it also has the forms needed to record investments and subsistence.

SETTING UP THE FIRST SESSION

( ) 2. Minority Group Option. Decide whether or not you wish to use the Minority Group Option, and announce whether it is in effect at the time you discuss the rules. Some discussion of this option is in order to give you a better basis for making a decision.

The model of minority groups employed in SIMSOC emphasizes one central characteristic of minority groups: a sense of precariousness. Minority group members are not deprived economically but are about as well off as non-minority group members, on the average. Some minority group members may even be heads of groups. Minority group members retain a vulnerability to arbitrary action by non-minority group members, even if no apparent notice is taken of their status and no minority-group action form is ever filed. The sense of precariousness may operate and affect the behavior of minority-group members, sometimes without awareness of this fact on the part of non-minority-group members. Differential perceptions of the importance of minority-group status frequently come to light in post-game discussion, with the non-minority-group members denying its importance and the minority-group members claiming continual consciousness of vulnerability. This is especially true among those who occupy highly visible positions of privilege or power in society. It is not uncommon for a minority-group member who is the head of a group to transfer this position to a trusted associate while he or she remains active behind the scenes.

Minority-group membership may be given varying degrees of salience by the insignia that are used to identify it. It is not a good idea to use insignia that too obviously suggest the experience of any particular minority group, since a sense of precariousness is common to minority groups in many different societies. For example, a yellow armband with a Star of David would be inappropriate because it too strongly suggests the treatment of Jews by the Nazis and can also be confused with the Yellow Region. The insignia should not overtly suggest to the participants that minority-group members are or should be discriminated against.

On the other hand, the insignia must be visible if minority-group status is going to have an opportunity to be salient. At a minimum, a minority-group member can be asked to wear an additional name tag marked MGM. A name tag of a distinctive color is even better as long as the color is not red, yellow, blue, or green, so as to avoid the confusion of minority group status with region.

You might wish to consider a range of more distinctive insignia. An armband, for example, is more distinctive than a name tag. (Again, be careful not to use a regional color.) A more dramatic way of identifying minority-group members is to provide them with eye masks that they must wear at all times. This is dramatically visible and has the added virtue of giving minority group members a salient physical feature that makes them look alike. It also shields their facial expressions, increasing the chance that others will feel less confident about what minority-group members are thinking, if this becomes an axis of conflict in the society.

(   ) 3. Size level of the society. Note the approximate number of participants and determine the size level using Table 1 below:

Table 1. Determining Size Level

Number of Participants	Size Level
15–32	One
33–47	Two
48–60	Three

Do not tell participants the exact population size. The taking of a census is an organizational problem that should be left to the participants to solve rather than to have it solved for them. However, it is appropriate to announce the overall size level during the discussion of the rules, if you are able to determine it.

(   ) 4. Listing participants on the Roster (Form Q). After you have collected the participants' Choice Sheets and Assignment Sheets, sort them into alphabetical order and list the names on the Roster. You may wish to number participants as well, to facilitate the use of assignments using random numbers.

(   ) 5. (OPTIONAL). Assigning people to the Red Region. If you know nothing of the participants, you can simply eliminate this step. Furthermore, even if you know them to some degree, you may prefer to leave all assignments completely to chance. Again, you would omit this step.

Many coordinators, however, prefer to intervene in the random process by introducing one or another aspect of purposeful assignment. Different purposeful assignments are possible, and you may wish to consider one or more of the following:

a.   Assignments based on prior measures of personality characteristics such as high need for power, ambitiousness, shyness, or the like. Some coordinators, interested in the interplay between personality and role, systematically vary the assignment between different SIMSOCs to explore the difference it makes. For example, they may assign those with high power-needs to head basic groups in one version while assigning those with low power-needs to these positions in another group of participants. I have never used such a method of assignment myself, and offer it here merely as an idea to consider.

b.   Assignments based on external status characteristics—especially sex and race (e.g., Glandon, 1973). Some coordinators have experimented with assigning women to head all groups or have varied the sex ratio in the Red Region in various ways. Again, this is an opportunity for the coordinator to experiment in interesting ways.

When I have used some version of this option, I have found it useful to reverse real-life roles. For example, those who occupy relatively powerful positions in real life would be given relatively powerless positions in SIMSOC and vice versa. There appears to be greater learning potential in role-reversal assignments than in reflecting outside status positions in SIM-SOC.

    c.   Stacking the Red Region. The one purposive assignment that I most frequently use is in the assignment of participants to the Red Region. The game works best when the Red Region is an active one. I try to place in it individuals who seem prepared to defend their interests in some active way. Whether they are likely to do this collectively or individually and their particular ideological predisposition are of less concern for me than their general inclination to some form of active effort to overcome their situation versus passivity and resignation. Of course, it is frequently hard to predict how participants will act in the game, and my assignments are sometimes based on intuitions that turn out to be wrong. No harm is done in the process, and the results of even error-prone guesses are likely to be at least as good as a random assignment in producing a vigorous Red Region.

If one chooses to adopt a non-random assignment, then this step involves choosing 25% of the participants for assignment to the Red Region, marking them appropriately on the Roster. (If you are running an undersized SIMSOC, choose 33% of the participants for assignment to the Red Region.)

( ) **6.** Assigning group heads. Sort the Choice Sheets (Form A) by first choice of a basic group. Taking each group in turn, select a head using either a random assignment or a purposive one as described above. If you have already made assignments to the Red Region, none of these participants is eligible to become a group head; eliminate these from the pile. Beyond that, use any criterion that suits your fancy. (Note: *Be sure to make Roster entries in pencil, since frequent changes are possible.*)

( ) **7.** Assign these group heads to regions as follows (if you have fewer than 24 participants, skip to the section called "Running an Undersized SIMSOC," p. 52, for steps 7, 8, 11, and 18):

    Green: Head of BASIN, JUDCO, and POP

    Yellow: Head of SOP and RETSIN

    Blue: Head of MASMED and EMPIN

    Red: No group heads

The head of MASMED has a Private Transportation Certificate which should be recorded on your Roster.

( ) **8.** Assigning subsistence and travel agencies. Using Table 2, determine the number of subsistence and travel agencies in the society.

Table 2. Number of Subsistence Agencies (S) and Travel Agencies (T) by Region

Number of Participants	Number of Agencies S	T	Green	Yellow	Blue	Red
24–28	4	2	2S/1T	1S/1T	1S/0T	0/0
29–35	5	2	3S/1T	1S/1T	1S/0T	0/0
36–40	6	2	3S/1T	2S/1T	1S/0T	0/0
41–45	7	3	3S/1T	2S/1T	2S/1T	0/0
46–52	8	3	4S/1T	2S/1T	2S/1T	0/0
53–57	9	3	4S/1T	3S/1T	2S/1T	0/0
58–60	10	3	4S/1T	3S/1T	3S/1T	0/0

Assign agencies to individuals at random (or according to some criterion of your choosing), making sure that no one gets more than one and that no group head or member of the Red Region gets one. Record these assignments on the Roster.

Table 2 provides an appropriate amount of scarcity. The proper amount of subsistence is a delicate matter. Too few subsistence tickets and the society may, in effect, be doomed from the outset; too many and the dynamics that flow from grappling with the problem of scarcity will not occur. The level set by Table 2 has some range of tolerance, but it should not be departed from radically.

Scarcity can be undercut by absenteeism, since absent members are not required to provide subsistence. An inadequate number of subsistence tickets may become adequate if enough people stay home. To counter this, the rules call for the elimination of tickets for one agency in any session in which five people are absent and the elimination of tickets for two agencies if ten people are absent. This should prevent a false prosperity through absenteeism. If there are fewer than five absentees in subsequent sessions, the full allotment of tickets is restored.

It is also possible for latecomers to join while the game is in progress. If there are as many as five latecomers in the game, you should add an additional subsistence agency to prevent scarcity from becoming too severe.

The amount of income provided in the game is sufficient to allow for full subsistence within two sessions by the purchase of enough luxury living to cover the subsistence gap. The gap can be eliminated in one session if a concentrated effort is made in this direction, but, even with a good proportion of Simbucks being used for other purposes, subsistence scarcity can be handled in two sessions. Travel scarcity can also become insignificant after a few sessions through the purchase of Private Transportation Certificates.

(   ) 9. Assign agency heads to regions so that the numbers per region conform to those specified in Table 2.

(   ) 10. Complete regional assignments. Assign all those who do not already have a specified region to some region, making sure that each region has approximately 25% (or 33% in an undersized SIMSOC). Again, you may use random assignment or assign according to some criterion of choice.

Table 3. Minority-Group Member Assignment by Region

Number of Participants	Green	Yellow	Blue	Red
24–27	1	3	1	0
28–32	1	4	1	0
33–37	1	4	1	1
38–42	1	5	1	1
43–47	1	6	1	1
48–52	1	7	1	1
53–57	1	7	2	1
58–60	1	8	2	1

( ) **11.** (OPTIONAL). Minority group assignment. Assign minority-group status to participants randomly within regions using Table 3 for the number per region, and record these assignments on the Roster.

The designated assignment has two intentions: to concentrate minority-group members in one region (Yellow), where they comprise about 50% of the population, and to make the general level of minority-group privilege approximately equal to that of non-minority-group members. It is useful to have a few group or agency heads with minority-group status.

( ) **12.** The Roster is now complete (see the sample SIMSOC Roster, p. 50, for an example of how one might look). You are now ready to prepare the participants' envelopes. Sort the envelopes in alphabetical order and place the following information on the outside of each one:

Upper left corner: Initial for region (R, Y, G, B).
Lower left corner: MGM if minority-group member, leave blank if not.
Lower right corner: Trav. or Sub. if agency head; otherwise leave blank.
Upper right corner: Name of basic group if head; otherwise leave blank.

Note that everyone will have a regional entry and that this may be the only entry for some individuals.

( ) **13.** Fill out the Assignment Sheets (Form B). Circle the appropriate information on this form, separate it from the Choice Sheet (Form A), and place it in the participant's envelope. You may dispose of the Assignment Sheets unless you wish to keep them as a record of individual goals.

( ) **14.** Name tags. Place a name tag in each envelope with any special insignia for minority-group members if this option is in effect.

( ) **15.** Agency tickets. Pull the envelopes of subsistence and travel agency heads and place five of the appropriate tickets in each envelope. Also pull the envelope of the head of MASMED and place five travel tickets in it.

( ) **16.** Starting income. Pull the envelopes of the basic group heads and give them their starting income (see Coordinator's Summary, Form P, Table 5, for the correct amount).

( ) **17.** Munchie tickets. Distribute Munchie tickets according to the following schedule:

SIMSOC Roster

Num‑ber	Name	Region	MGM	Certificates and Agencies	B	R	P	S	E	M	J	1	2	3	4	5	6	7	8	9	10
1	Albert Adler	Y		T																	
2	Bernie Berger	R																			
3	Chuck Clark	G					H														
4	Dave Duncan	B																			
5	Elvis Edwards	Y					H														
6	Frank Foster	G						H													
7	Gino Geribaldi	R																			
8	Howie Hyman	G		S																	
9	Jonathan Jackson	Y	✓	S																	
10	Ken Kirby	G	✓	T																	
11	Larry Lowenthal	R																			
12	Mel Myerson	B		S																	
13	Oscar Overseth	G									H											
14	Peter Potter	Y	✓																		
15	Sam Stryker	G		S																	
16	Tom Tusker	R																			
17	Vic Volstead	Y	✓						H												
18	Alice Burrows	B																			
19	Brenda Carson	R																			
20	Dorothy Epstein	B	✓																		
21	Ellie Felch	R																			
22	Francis Gentry	G		S																	
23	Ginny Horst	Y	✓																		
24	Helen Jones	R																			
25	Jean Kelly	B							H												
26	Kate Mallory	B		P						H											
27	Nadia Orfelder	Y																			
28	Pat Regent	B																			
29	Sally Thomas	G																			

KEY

Region
R = Red
Y = Yellow
G = Green
B = Blue

MGM
Check if person is a minority‑group member.

Agencies/Certificates
S = Subsistence
T = Travel
L = Luxury Living Endowment
P = Private transportation

Basic Group
B = BASIN
R = RETSIN
P = POP
S = SOP
E = EMPIN
M = MASMED
J = JUDCO

✓ = Employee
H = Head

Subsistence
S = Provided
X = None provided
A = Absent
D = Dead
J = Jailed (under arrest)

Table 4. Initial Munchie Distribution

| Size Level | Hold in Bank | Group Heads (5 per head) | Randomly to Individuals in: | | | | Total |
			Green	Yellow	Blue	Red	
One	35	(7 × 5) = 35	0	0	0	0	70
Two	50	(7 × 5) = 35	3	3	3	3/3	100
Three	65	(7 × 5) = 35	5	5/5	5	5/5	130

The schedule calls for 50% of the available tickets to be held for sale by the coordinator. These should be put aside in a special envelope, marked for this purpose. Each group head receives five tickets. In a Size Level Two society, the remaining 15 tickets are given out, in lots of three, to five individuals in the designated regions, while in a Size Level Three society, the remaining 30 tickets are given out, in lots of five, to six individuals in the designated regions.

(   ) 18. Sort the envelopes by region and fill out the number of subsistence and travel agencies on the four copies of the Regional Summary Sheet (Form R).

(   ) 19. Set up a Munchie Bazaar. The coordinator may need to experiment somewhat with the precise materials to include in the Munchie Bazaar depending on the circumstances under which the game is being run. Certain general principles should be kept in mind in setting it up:

a.   Munchies should not be so attractive that they lure Simbucks from those who are committed to societal or group goals. To take an extreme case, if participants are playing in a hot, thirst-producing environment and the Munchie Bazaar offers cold beer for a few Simbucks, even the most committed members of the society may feel driven by thirst to use their Simbucks for personal consumption. Munchies should represent luxuries that can be foregone easily enough by individuals who want to use their Simbucks for public purposes rather than private consumption.

b.   Munchies should be attractive enough to offer an alternative means of using Simbucks for individuals who do not care much about societal or group goals. The willingness to purchase Munchies should reflect the changes and variations among individuals in their willingness to forego personal consumption in order to achieve social purposes. If Munchies have too little attraction, even an individual with very low commitment to social purposes may feel that he has little use for his Simbucks. Thus, he may invest them in Public Programs for want of any viable alternative instead of through some positive choice to do so. Munchies should present a private consumption alternative to those with low commitment.

c.   Do not be disturbed if the Munchie Bazaar gets only limited use, especially during the early phases. The model underlying the game suggests that Munchies will be used at the beginning mainly insofar as they come to participants through their purchase of Luxury Living Endowments or "inherited" Munchie Tickets. During the period of great scarcity, the purchase of Munchie Tickets directly from the coordinator should be relatively uncommon. At the end of Phase Two and the beginning of Phase Three, the purchase of Munchie Tickets should be more common if they are still available in the society, and they should more frequently be purchased by non-privileged members of the society who have managed to acquire some money.

d.   Suggested Munchie price list. Because demand tends to shift sharply during the three phases of the game, a three-tiered price structure is suggested. It is listed below in terms of phases of the game, but you will want to time price changes to the actual supply and demand in the game you are running. The following suggested list prices should give you a good place to start:

Table 5. Suggested Munchie Prices

Item	Phase One	Phase Two	Phase Three
Chocolate chip cookie, pretzel	1M*	2M	3M
Candy bar, doughnut, small bag of potato chips or peanuts	5M	7M	10M
Beer, soft drink	8M	12M	20M

* M = Munchie Ticket.

e.   The actual dollar cost of such refreshments may be passed on to the participants by charging real money in addition to Munchie tickets. You may wish to do this only for more expensive items such as beer and soft drinks. Note, however, that the total number of Munchie tickets available to the society will be limited, so the coordinator may want to simplify matters by absorbing the actual cost.

f.   You may want to decorate your Munchie Bazaar with a colorful canopy, posters, and appropriate signs.

RUNNING AN UNDERSIZED SIMSOC

If you have fewer than 24 participants, certain adjustments must be made. The main reason for these adjustments is that the game is not as effective if regions are too small. Thus, the main change is to eliminate one of the regions. Steps 1 through 6 are the same, but step 7 must be altered.

(   )**7U.** (Use with 23 or fewer participants.) Assign the group heads to regions as follows:

Green: Head of BASIN, JUDCO, POP, and EMPIN.

Blue: (This region does not exist in an undersized SIMSOC.)

Yellow: Head of MASMED, RETSIN, and SOP.

Red: No group heads.

(   )8U. Assign agency heads according to Table 6:

Table 6. Number of Subsistence Agencies (S) and Travel Agencies (T) by Region for Undersized SIMSOCS

Number of Participants	Number of Agencies			Region	
	S	T	Green	Yellow	Red
15–18	2	1	1S/1T	1S/0T	0/0
19–23	3	1	2S/1T	1S/0T	0/0

Waive the rule that no more than one-third of all the participants may live in any one region. The rule prohibiting more than 50% from being present in any one region remains in effect.

Steps 9 and 10 remain the same.

(   )11U. Assign minority-group members as follows: Green = 1, Yellow = 3, Red = 0.

Steps 12 to 17 and step 19 remain the same.

(   )18U. Correct the Regional Summary Sheet (Form R) as follows:

Cross out the Blue Region and add EMPIN to Green and MASMED to Yellow. Then add the appropriate number of subsistence and travel agencies by region.

SIMSOC Playing the Game

THE FIRST SESSION

(   )20. Before the participants arrive, you should:

a. Sort the envelopes by region and place them in the appropriate region;

b. Set up the Munchie Bazaar;

c. Set up the needed materials for the head of MASMED.

As participants arrive, consult your Roster and send them to their home region, informing them that they have an envelope awaiting them there. It is not necessary to wait for everyone to arrive to begin. Once about two-thirds of the participants have arrived, you may allow them to begin, with late arrivals simply joining in as they get there. Do not reassign any group heads or agencies until at least twenty minutes into the session, when you are quite sure the person is absent.

( )21. Once the game has begun, the coordinator or his assistants should go to each region and:

a.   Leave the Regional Summary Sheet (Form R) for the region;

b.   Remind everyone to wear his or her name tag at all times. It is especially important for minority-group members to wear their insignia so that they can be identified at all times;

c.   Remind the heads of BASIN and RETSIN that they can purchase passages or anagrams at any time during the session, but must complete their task during the session of purchase to receive credit; make sure that the head of MASMED understands the options available for broadcasting or publishing a news magazine;

d.   Remind all group heads that they will need to file a Job Schedule (Form G) during the first session, so that the coordinator knows who is officially employed. Group heads should understand that it is not necessary for them to file this form after the first session unless they wish to make changes; employment carries over from session to session automatically unless a new form is filed.

( )22. An assistant should remain at the coordinator's desk at all times to handle transactions, man the Munchie Bazaar, and collect travel tickets. It is best to collect travel tickets at the beginning of a trip. The coordinator should remain mobile so he can handle inquiries and record information as needed.

( )23. With a half-hour or so remaining in the session, the coordinator or an assistant should visit each region with the following announcement:

a.   Announce the amount of time remaining in the session;

b.   Ask if anyone would like to provide for his or her subsistence. Collect any offered subsistence tickets and record this information on the Roster;

c.   Ask if anyone would like to invest in Public Programs, and record any investments on the Investment Tally Sheet (Form V). It is very unusual for any investment in Public Programs to take place in the first session, but this is a way of reminding participants of the existence of this option. (See the sample Investment Tally Sheet, p. 55, for an example of how this form might look after a couple of sessions. Note that it is necessary to record only the amount, not the name of the donor.)

d.   Ask if anyone would like to turn in a political party support card, an EMPIN membership card, or a MASMED subscription. Make it clear that this is an option that they are free to exercise or not. This reminder is intended to make them aware of the option without exerting any pressure on them to exercise it. Collect all group support cards that are handed in, since

Investment Tally Sheet

Form V

(Sample for Size Level Two SIMSOC)

Session	BASIN Passages						RETSIN Anagrams						Public Programs	
	1	2	3	4	5	Total	1	2	3	4	5	Total	R&C [a]	WS [b]
1.	69	75				144	72	90				162	0	0
2.	63	75	69	75		282	72	90	72			234	0	2,1,1= 4
3.														
4.														
5.														
6.														
7.														
8.														
9.														
10.														

SAMPLE

Payment Schedule		Size Level	
	One	Two	Three
BASIN (per passage)	50	75	100
BASIN error deduction (per error)	4	6	8
RETSIN (per marketable word)	12	18	24

a R&C = Research and Conservation
b WS = Welfare Services

you will want to sort these when you return to the coordinator's table.

e.   Ask if anyone would like to turn in an individual goal declaration card. Again, it should be made clear that this is an option that they need not exercise.

(   ) 24. About five minutes before the end of the session, the coordinator or an assistant should visit each region and announce that the session is ending in five minutes and that this is the last opportunity they will have to turn in subsistence tickets, support cards, investments, and the like. He should collect whatever is given him, recording subsistence on the Roster and taking support cards back to the coordinator's table. I generally accept cards that are brought to the coordinators table at the last minute, until I have declared that the session is officially over and have begun tabulating the results.

There is a tendency for the coordinator's job to become quite hectic at the end of a session. If too many people crowd around the coordinator's desk at the end of the session, ask everyone to return to his or her home region and send someone to the respective regions to handle the final transactions. In general, it is desirable to keep the number of players at the coordinator's table down to one or two persons, and to handle questions away from the table.

(   ) 25. Collecting envelopes. Be sure to collect the envelopes of all participants at the end of each session. You will need them to distribute materials for the following session.

THE INTERSESSION

(   ) 26. Setting up for the intersession. It is desirable to prepare for the next session right after the previous session. If another session will follow shortly, you have no choice on this, but it is best to complete the task right away even if the next session does not follow immediately. If participants remain for another session, ask them to stay in their home regions (except to go to the bathroom and the like) and not to talk to members of other regions during the intersession. Keep them away from the coordinator's table and ask them not to interrupt you, since you will have your hands full preparing for the next session.

The preparations between sessions can be divided into three major tasks: *calculating basic group income, calculating the National Indicators,* and *preparing the envelopes for the next session.* The first two of these tasks can occur more or less simultaneously.

(   ) 27. Form X: Basic Group Income Calculations. It is useful to train an assistant in this task and to set aside a separate coordinator's table for all the relevant materials. The person performing this task should have the following:

a.   All Withdrawal of Asset forms (Form I) filed during the session. These are the "checks" that the BASIN and RETSIN heads issue during the session. Once this information is recorded, dispose of these slips so that they will not become mixed in with such forms from later sessions. The amount withdrawn is filled in on line 2 of Form X.

b. POP and SOP support cards, EMPIN membership cards, and MASMED subscriptions. Whenever these have been collected during the session, they should be placed on the table where group income calculations will be made. These cards need to be sorted and counted. Be sure to remove EMPIN cards for unemployed people and all cards from individuals who are under arrest. The valid cards remaining should be counted and the numbers entered on line 5, Form X.

c. The person performing these calculations needs to obtain from Form V (the Investment Tally Sheet) the payment that BASIN and RETSIN have coming to them. This total for each should be recorded on line 3, Form X.

d. The National Indicator multiplier (line 11, Form X) will not be available until the National Indicator calculations have been completed. Form X can be completed through line 10 while awaiting this final piece of information.

e. Once the National Indicator multiplier is available, line 12 can be completed. The person in charge of Form X should then place the appropriate amount of Simbucks called for in line 12 in the envelopes of the heads of groups.

(For an example of a filled-out Basic Group Income Calculations form, see p. 58.)

( ) 28. National Indicator Calculations (Form W). The coordinator or another assistant should take regular responsibility for this set of calculations. (Round off all calculations to the nearest integer.)

a. The information on investments called for in lines 3–6 is available from the Investment Tally Sheet (Form V).

b. Number of absentees is available from the Roster.

c. To calculate the number of unemployed, you must do two things:

(i) Record the information on who has been hired from the seven Job Schedules (Form G) that have been filed by group heads. If these forms were turned in during the session, you may have had the opportunity to do this before the end of the session, which will make your job much easier at this point. It is not desirable that unemployment occur simply because a group head has forgotten to turn in his form. Consequently, I remind any group head who has failed to file Form G to do so until I am confident that the failure to file a Job Schedule is an act of choice rather than an oversight.

(ii) Note who has failed to provide subsistence and remove them from any jobs which they may have been listed for. They are counted as unemployed.

Now you are ready to count the number to be recorded on line 8 of Form W.

Basic Group Income Calculations

(Sample for Size Level Two SIMSOC)

Form X

Session # _____

	BASIN	RETSIN	POP	SOP	EMPIN	MASMED	JUDCO
1. Assets at beginning of session	144	181					
2. Assets withdrawn on Form I	-144	-180					
3. Payments due for product (Totals from Form V)	+282	+234					
4. Net assets in bank	282	235					
5. Support cards turned in			22	8	27	30	
6. Starting income [a]			60	60			
7. Support product (multiply line 5 by line 6)			1320	480			
8. Total population (TP) = __40__ (divide line 7 by TP)			33	12			
9. Support multiplier			2.5	2.5	2	2	
10. Basic income [b]	28	24	83	30	54	60	45
11. National Indicator multiplier [c] (__30__%)							
12. Net Income (line 10 × line 11)	8	7	25	9	16	18	14

[a] For POP & SOP: Size Level One = $40
Two = $60
Three = $80

[b] Basic Income:
For BASIN/RETSIN = 10% of line 4
For POP/SOP = line 8 × line 9
For EMPIN/MASMED = line 5 × line 9
For JUDCO = Size Level One = $30
Two = $45
Three = $60

[c] Use new National Indicator levels from Form W and Table 2 from Form P to obtain correct figure.

> **d.**   Number of rioters, arrests, and deaths should be readily obtainable from your records.

> **e.**   Sort the individual goal declarations into positives and negatives. Be sure to remove the goal declarations of anyone under arrest. Note that line 12 calls for a net figure which is 25% of the positive declarations minus 100% of the negatives. Circle the appropriate plus or minus sign under the Public Commitment (PC) column of line 12.

Once you have calculated the final figures, check Table 2 of the Coordinator's Summary (Form P) to get the National Indicator multiplier. This information should be given to the person who is calculating group incomes. (An example of a filled-out National Indicator Calculations form is provided on p. 60.)

( ) **29.** You are now ready to prepare the envelopes for the next session.

> **a.**   If it has not already been done, pull the envelopes of group heads and give them the Simbucks to which they are entitled.

> **b.**   Pull the envelopes of subsistence and travel agency heads, and give each person five of the appropriate tickets.

> **c.**   Count out the appropriate number of Munchie tickets that will be available for the upcoming session. Note that this depends on the size level of the society and the level of the Standard of Living. Once you have collected the appropriate total, distribute five Munchie tickets to each person who has purchased a Luxury Living Endowment. It will be convenient to make some mark indicating this luxury-living status on their envelopes to facilitate this operation in the future.

> **d.**   Fill out the Report to MASMED (Form Y) and place it in the envelope of the MASMED head.

> **e.**   Re-sort the envelopes by region. If participants are waiting for the next session to begin, return the set of envelopes to the appropriate regions and allow them to begin the next session.

LATER SESSIONS

( ) **30.** Reminders of a general sort are no longer necessary. However, you should inform the heads of BASIN and RETSIN of their current assets in the bank. If anyone is under arrest, ask the head of the arresting Simforce whether renewal of the arrest is desired and collect if it is. Check whether anyone who has posted a guard in a region wishes to renew. If a Simforce is due for renewal, check with the head whether he or she desires to renew. If the number of absentees has risen to more than five, remove one subsistence agency allotment at random.

( ) **31.** The coordinator's life will be simpler if participants hand in subsistence and other materials during the session, instead of waiting until the rush at the end. It is worthwhile to make one round of the regions during the session for the purpose of collecting such materials.

National Indicator Calculations

Session # ___2___

National Indicators

	FES	SL	SC	PC
1. Initial value (at beginning of session)	86	68	72	78
2. Natural decline (10% of line 1)	– 9	– 7	– 7	– 8
3. Research and Conservation = ___0___ (FES = +40%; SL = +10%)	+ 0	+ 0	0	0
4. Welfare Services = ___4___ (SL = +10%; SC, PC = +20%)	0	+ 0	+ 1	+ 1
5. BASIN passages bought = ___4___ (FES = –2; SL = +1)	– 8	+ 4	0	0
6. RETSIN anagrams bought = ___3___ (SL = +1; PC = –1)	0	+ 3	0	– 3
7. Number of absentees (Ab) = ___2___ (SL, PC = –2Ab)	0	– 4	0	– 4
8. Number of unemployed (U) = ___5___ (SL, SC = –3U; PC = –1U)	0	–15	–15	– 5
9. Number of rioters (R) = ___10___ (For SC, see Form P; PC = –2R)	0	0	–20	–20
10. Number of arrests (Ar) = ___2___ (SC, PC = –3Ar)	0	0	– 6	– 6
11. Number of deaths (D) = ___1___ (SL, SC, PC = –5D)	0	– 5	– 5	– 5
12. Individual goal declarations: Positive (P) = ___26___ ; Negative (N) = ___9___ (Pt = .25P-N)	0	0	0	±3
13. Special events	0	0	0	0
14. Total of all pluses	+ 0	+ 7	+ 1	+ 1
15. Total of all minuses	–17	–31	–53	–54
16. Net change (add lines 14 & 15)	–17	– 24	–52	–53
17. Final value (add lines 1 & 16)	69	44	20	25

SAMPLE

SPECIAL EVENTS

( )**32.** You may wish to introduce special events in the course of the society to enliven things or to give a society on the verge of collapse a chance to recover. Sometimes activities develop slowly and build to a climax, and the coordinator should not be too eager to introduce disturbances. In many cases, it may be wiser to let the society run its natural course without introducing any special events. Five events are discussed below; they are described in full detail on detachable forms at the back of this manual.

a.   *Massive Foreign Aid.* Benjamin Gorman[*] suggested this idea for societies in which the National Indicators have declined to the point of collapse. It presents SIMSOC with a chance to boost most of the National Indicators by 50 points if the players are willing to relinquish their autonomy to a foreign power. In one case, the conditions involve surrendering the industries, BASIN and RETSIN, to foreign ownership. In a second case, they involve giving the Red Region control of all of the basic groups in the society. In a third case, it involves action against minority-group members, including some arrests.[†] Acceptance of these offers is voluntary and must be with the consent of a majority of the members as long as the National Indicators are above zero. If, however, a National Indicator has fallen below zero and you wish to continue the game, you may require that at least one of the offers be accepted. The offers are compatible with each other, and it is possible to accept all three.

b.   *Expeditionary Force.* This is another possibility for a society in which the National Indicators are very low. It offers an opportunity for a rapid recovery of the National Indicators by means of a successful foreign adventure. An expeditionary force does not make sense strictly from the standpoint of raising National Indicators. Its use is as likely to lower as to raise the National Indicators, by an equal amount. The same amount of money invested in Research and Conservation and Welfare Services would be certain to raise the National Indicators (although by a much more modest amount). Under these circumstances it should be less of a temptation for a society with National Indicators in good shape.

c.   *Epidemic in the Red Region.* This provides the society with two kinds of problems: keeping the epidemic from spreading and keeping individuals from dying. If members of Red have the means to travel, the epidemic cannot be prevented from spreading without their active cooperation. If the epidemic is contained and members of other regions allow members of Red to die, the National Indicators will be affected. Investing the money in Welfare Services rather than supporting immunization is, however, a cheaper method of supporting the National Indicators. In sum, an epidemic requires some organizational skill

[*] Benjamin Gorman is a sociologist at the University of Florida.
[†] This idea is based on a suggestion by Charles G. Waugh.

to handle, and, because only one subgroup is affected, it presents members with some issues about how much responsibility they should take for the welfare of others.

d.   *Earthquake*. The earthquake cuts the society in half, allowing travel within either half but not from one half to the other. Because travel is the major means of communication, this event complicates many problems of coordination. Furthermore, the expense of repairing the damage is great enough that there may be problems in poorly organized societies with raising the money. Which half, for example, will pay for the repairs? If and when the damage has been repaired, the coordinator should announce it immediately and return travel conditions to their normal state.

e.   *Foreign Threat*. This is one of three events that may have important effects on the National Indicators. The effects are set in such a way that it is most rational to take no action if keeping the National Indicators up is the only goal. (Players should not be told this, of course.) If the amount of money invested in a defense force were invested equally in Research and Conservation and in Welfare Services, the loss from the foreign threat would be completely offset. This would have essentially the same effect as a successful defense force, but such a force has only an 0.8 probability of being successful. (The slight gains in Social Cohesion and Public Commitment from a successful defense force are balanced by the risk of a substantially larger loss if the force is unsuccessful.) Players may, however, invoke other goals, such as national honor and pride, and the decision on how to respond can become complex. If the society has established no clear decision-making procedure and structures, this event may stimulate such development or lead to changes in existing procedures.

You may wish to use anywhere from none to all five of these special events. Furthermore, you may want to add others of your own devising. But if you do the latter, follow the same procedure recommended here: Introduce them as "events" which do not alter any of the existing rules. Attempts to change the basic rules of the manual while the game is in progress generally meet with resistance as an arbitrary intervention by the coordinator. It is probably wisest not to introduce any events, allowing the society to follow its natural course.

A Special Events Sheet (Form Z) is provided to help you keep a record of the response to these events should you decide to use one or more.

THE FINAL SESSION

The coordinator must judge the appropriate time to end the game. Certain guidelines will be helpful in avoiding end-game effects. As noted earlier, it is a good idea to end the game in the middle of a session rather than at the end. It is best if the participants anticipate that there will be at least one additional session at the time the game is ended. Indications that the game is moving into

Phase Three are a useful signal to the coordinator that the game has run its course (see pp. 22–23).

Participants have sometimes suggested that it would be useful to have a discussion while the game is in progress, relating what was happening to course readings and events in the world. I have experimented with this in a very limited fashion and have found one problem: it is difficult to keep the discussants from considering how to play the game. Many are still interested in influencing what happens in SIMSOC rather than in analyzing it. When this occurs, the discussion becomes a SIMSOC meeting of the committee-of-the-whole rather than a group of social analysts attempting to make sense of their experiences. For this reason, I believe it is more desirable to defer all discussion until the conclusion of the game.

SIMSOC The Post-Game Discussion

This is the heart of the learning experience. Along with written work based on the game, it is the *raison d'être* for SIMSOC, and sufficient time should be reserved so that it can be carried on fully. The art of the post-game discussion is to crēate an atmosphere in which the participants are talking to you and their fellow participants rather than listening to you talk to them. It is very easy for people to fall into a pattern of listening to the coordinator and asking questions instead of having the coordinator ask the questions while the participants make the observations.

Many SIMSOC users have found the following procedure for the post-game discussion a helpful one. Think of the post-game discussion as divided into three parts: catharsis, participants' analysis, and coordinator's analysis.

CATHARSIS

Participants are still quite involved in their roles in the game when the discussion begins. Their comments at the beginning do not have a very detached, analytic quality; usually, they are thinly disguised justifications for their own actions in the game. Frequently they imply criticism of other participants, who will then respond defensively, attempting to justify the actions being criticized. It's impossible to prevent a certain amount of these self-justifying comments, but I usually try to discourage them and to move the discussion toward a greater degree of detachment.

One device that helps is to allow members of only one region at a time to speak during the first phase. Typically, I start with the Red Region and then choose the other regions in turn, depending on their salience as protagonists in the society. A good starting question is to ask the members of each region to describe how they felt at different points in the game. If these feelings are dealt with and discussed, it will be easier to move to the second, more analytic phase of the game. If participants from other regions react defensively and attempt to interrupt, ask them to wait their turn. It sometimes helps to make such comments as: "Instead of arguing, try to listen to what he is saying for a minute. He is trying to tell you how he saw things as a member of the Red Region. Ask yourself why something might have looked different to him, coming from that region. Don't worry about convincing him of your version of the truth." Gradually, as members of each region chime in with their version of events, the tone of self-justification should disappear and a more analytic tone should emerge.

The discussion questions suggested for this state are as follows:

1. Please describe how you felt at the beginning of the game and how your feelings changed during the course of it.

2. What would you consider the critical events in the life of the society, and how did you respond to them when they occurred?

PARTICIPANTS' ANALYSIS

During this phase of the discussion, the objective is to get participants to discuss the analogies between processes that occurred in SIMSOC and those in the real world. At this point, the constraint of people speaking by region can be abandoned. By now, participants should have abandoned their game roles in favor of the role of social analyst. It is best for the coordinator to withhold his or her own observations until the final phase of the discussion.

You may wish to start this phase with the following general questions:

1. Was there anything that happened in the course of the society that reminded you of things you have observed in the real world? If so, please describe it.

2. Was there anything that happened in the course of the society that seemed quite different from the way things happen in the real world?

These questions can be repeated several times, encouraging additional people to make observations. It is useful to ask for agreement or disagreement from other participants about the observations that are made and to encourage discussion among participants on the accuracy of the analogies.

You may also wish to prepare a series of questions about specific events in SIMSOC which you think will highlight more general issues. For example, "Why do people think that Marjorie couldn't get the society to support her investment plan, even though she had a Simforce and no one else did?" "Why do you think no Simforce was ever created?" "Why do you think the society almost collapsed at the beginning?" Some of the study questions listed in the *Participant's Manual* (p. 33) may be appropriate for this purpose. A very small number of questions are sufficient for the purposes of most post-game discussions.

COORDINATOR'S OBSERVATIONS

In a good post-game discussion, the participants will discover for themselves many of the processes that SIMSOC is designed to highlight. However, you will undoubtedly notice aspects of what has been occurring that the participants have missed. This final phase is your opportunity to point out whatever you think they may have overlooked and, perhaps, to introduce a more abstract analysis of the processes they have been discussing.

This is the point at which to introduce the model underlying SIMSOC (pp. 14–24) if you find it useful to do so. It is easily detachable in the event that you wish to photocopy it for distribution to participants.* Or you may have your own analysis of what you have observed in this particular SIMSOC. Perhaps it does not fit the model very well, and you wish to note and make sense of the departures. If you have criticism of the validity of the model or its expression in the game, they should be shared with participants during this phase. If you disagree with observations made by some of the participants, this is your opportunity to suggest alternative interpretations which you withheld during the earlier discussion.

Note: Evaluating participants on the basis of their contributions to post-game discussion is not generally a good idea, since it will greatly inhibit discussion and exploration. If some evaluation of learning is going to occur, it is best to use written work for this purpose, using the post-game discussion to aid the participants in analyzing their experiences.

OTHER TYPES OF POST-GAME DISCUSSION

The description above focuses on the use of SIMSOC to gain insight into processes of large-scale conflict, protest, social control, and social change. For those using it to explore interpersonal feelings, communication, trust, and other aspects of face-to-face interaction, some of the previous advice is irrelevant. The catharsis state, instead of being viewed as an obstacle to analysis to be gotten out of the way, becomes the heart of the discussion. The object of the discussion leader is to bring people to self-insight or a better understanding of how they affect others in interpersonal interaction. The analogy to processes in the real world is less relevant since, for this purpose, SIMSOC is the real world.

Similarly, if the purpose is to allow students to explore the challenges of creating utopia, this should be reflected in a different emphasis in the post-game discussion. The central questions to explore are the reasons for any difficulties or failures in achieving utopia. The simulation aspects become relevant again because it is important to ask whether these obstacles are artificial ones created by the game or a reasonable reflection of those that exist in real life.

* Permission is hereby granted for photocopying pp. 14–24.

SIMSOC Written Work

The post-game discussion is an important step in turning the experience of playing SIMSOC into a learning experience. A second step involves written work or other projects based on SIMSOC. Some suggestions follow:

1. Have the participants write papers addressing issues that are made more vivid by their participation in SIMSOC. Assignment one in the *Participant's Manual* suggests a series of questions that players should be stimulated to think about by participating in SIMSOC. You will undoubtedly think of others based on what happened in a particular SIMSOC. The objective of these questions is to move the focus from SIMSOC to the processes in the real world that the game attempts to simulate. These questions can be used for further class discussions as well as for written work.

2. Ask the participants to become simulators. Playing a game may be a more active experience than listening to a lecture, but developing a game is more active still. Assignments two and three in the *Participant's Manual* are attempts to have students make the jump from player to simulator. As players, they accept certain constraints which are given in the rules and use resources to achieve objectives specified or suggested by the rules. When they act as simulators, the rules themselves become the resources, and they can manipulate these—hypothetically or actually—to see whether the resulting process will take the form that they believe it will. Depending on the emphasis you wish to give this aspect of your course or program, you may wish to assign some of the readings listed in the brief game-simulation bibliography included in this manual (p. 67).

3. Ask participants to keep a diary while they are playing. If the focus in using SIMSOC is on interpersonal behavior, these diaries may be a rich source of material to use later, especially if you ask people to record their feelings about what is going on rather than simply to describe events. The diaries can help the participants to recall reactions that may have become distorted and "corrected" by hindsight. Clovis R. Shepherd and colleagues at the Department of Sociology, University of Cincinnati, Cincinnati, Ohio, have developed a form for such a diary.

4. Have participants do research on SIMSOC. Lawrence Alschuler at the University of Hawaii has developed a research exercise based on SIMSOC. Some students, instead of playing, carry on systematic research and evaluation, developing questionnaires or special observation techniques, interviewing participants, and analyzing and reporting results. It is possible to carry out replications of existing studies (for example, surveys about attitudes toward government and politics) using SIMSOC's government and politics as the reference point. Questions and attitude scales used in other studies can be reworded for this purpose. Alschuler discusses this idea more fully in an unpublished paper, "Simulating Politics and Simulating Research: A Teaching Idea" (1969).

SIMSOC References and Bibliography

ON GAME SIMULATIONS

The following list includes books and articles referred to in this manual plus a number of other items for those who want to pursue game simulations further.

Abt, Clark. *Serious Games.* New York: Viking Press, 1970.

Adair, Charles A., and Foster, John T., Jr. *A Guide for Simulation Design.* Tallahassee, Florida: Instructional Simulation Design, 1972.

Avedon, Elliott M., and Sutton-Smith, Brian. *The Study of Games.* John Wiley, 1971.

Boocock, Sarane S., and Schild, E. O. *Simulation Games in Learning.* Beverly Hills, California: Sage Publications, 1968.

Caillois, R. *Man, Play, and Games.* New York: The Free Press, 1961.

Campbell, Donald T. "Pattern Matching as an Essential in Distal Knowing." In *The Psychology of Egon Brunswik,* edited by Kenneth R. Hammond. New York: Holt, Rinehart and Winston, 1966.

Coleman, James S. "Introduction: In Defense of Games." *American Behavioral Scientist,* vol. 10 (October, 1966), pp. 3–4.

Coppard, Larry C., and Goodman, Frederick, eds. *Urban Gaming/Simulations '77.* Ann Arbor: University of Michigan Publications Service, 1977.

Duke, Richard D. *Gaming: The Future's Language.* John Wiley, Halsted Press, 1974.

Guetzkow, Harold, ed. *Simulation in the Social Sciences.* Englewood Cliffs: Prentice-Hall, 1962.

Guetzkow, Harold; Alger, Chadwick F.; Brody, Richard A.; Noel, Robert C.; and Snyder, Richard C. *Simulation in International Relations: Developments for Research and Teaching.* Englewood Cliffs, N.J.: Prentice-Hall, 1963.

Gillespie, Philip. *Learning through Simulation Games.* New York: Paulist Press, 1973.

Gordon, Alice Kaplan. *Games for Growth.* Palo Alto, California: Science Research Associates, 1970.

Greenblat, Cathy S., and Duke, Richard D., eds. *Gaming-Simulation: Rationale, Design, and Application.* John Wiley, Halsted Press, 1975.

Greenblat, Cathy S.; Stein, Peter J.; and Washburn, Norman E. *The Marriage Game.* New York: Random House, 1974.

Horn, Robert E. *The Guide to Simulation Games for Education and Training.* Lexington, Mass.: Information Resources. Periodically revised and updated—write Didactic Systems Inc., Box 457, Cranford, N.J. 07016.

Huizinga, J. *Homo Ludens.* Boston: Beacon Press, 1962.

Inbar, Michael, and Stoll, Clarice, eds. *Simulation and Gaming in Social Science.* New York: The Free Press, 1972.

Krupar, Karen R. *Communication Games.* New York: The Free Press, 1973.

Lauffer, Armand. *The Aim of the Game.* New York: Game Simulations, Inc., 1973.

Livingston, Samuel, and Stoll, Clarice S. *Simulation Games for the Social Studies Teacher.* New York: The Free Press, 1973.

Raser, John R. *Simulation and Society.* Boston: Allyn and Bacon, 1969.

Sackson, Sid. *A Gamut of Games.* New York: Castle Books, 1969.

Stadsklev, Ronald. *Handbook of Simulation/Gaming in Social Education.* Birmingham, Ala.: Institute for Higher Education Research and Services, 1974.

ON SIMSOC

Alschuler, Lawrence. "Simulating Polities and Simulating Research: A Teaching Idea." Honolulu: University of Hawaii (Dept. of Political Science), 1969.

Dukes, Richard L., and Waller, Suzan J. "Towards a General Evaluation Model for Simulation Games: GEM." *Simulation and Games,* vol. 7, no. 1 (March, 1976), pp. 75–96.

Gamson, William A. "SIMSOC: Establishing Social Order in a Simulated Society," *Simulation and Games,* vol. 2 (September, 1971), pp. 287–308.

Gamson, William A., and Stambaugh, Russell J. "The Model Underlying SIMSOC." Working Paper #152, Center for Research on Social Organization, University of Michigan, 1977.

Gessner, John C. (Review of SIMSOC.) *Teaching Sociology,* vol. 2 (April, 1975), pp. 225–226.

Glandon, Nancy D. "Sexism: Rigging SIMSOC to Make the Point." Redlands, Calif.: University of Redlands, 1973. Available from author, c/o Division of Social Science, University of Redlands, Redlands, California 92373.

Green, Charles S. III. "Scarcity and the Distribution of Power. Some Insights from SIMSOC." Charlottesville, Va.: University of Virginia (Department of Sociology), 1975.

Nikkel, Stan R. "A Review of Urban Instructional Simulations." *Simulation and Games,* vol. 7, no. 1 (March, 1976), pp. 97–106.

Rosen, Bensen; Jerdee, Thomas H.; and Hegarty, W. Harvey. "Effects of Participation in a Simulated Society on Attitudes of Business Students." *Journal of Applied Psychology,* vol. 57 (1973), pp. 355–357.

Silver, Burton R. "Social Mobility and Intergroup Antagonism: A Simulation." *Journal of Conflict Resolution,* vol. 17 (December, 1973), pp. 605–623.

Stern, Robert N. "Market Behavior in a Simulated Society: Some Problems in Economy and Society." Nashville, Tenn.: Vanderbilt University. Available from author, Department of Sociology, Box 1811, Station B., Vanderbilt University, Nashville, Tennessee 37235.

Wolf, C. P. "SIMCANSOC: Simulated Canadian Society." *Simulation and Games,* vol. 3 (March, 1972), pp. 53–77.

SIMSOC Appendix: Table of Random Digits

A convenient way of using the Table of Random Digits is to pick any horizontal row as a start before looking at the table. Then, beginning at the extreme left, proceed forward, crossing out digits as you use them. When you need to use the table again later, you may simply continue from the previous point. Numbers which are not associated with players (for example, number 46 and above when there are only 45 participants) should be treated as blanks and should also be crossed off as you proceed.

54846	13302	37944	12648	13534	75987	17651	53363	59762	14154
60721	87738	63009	64254	25610	09488	27219	54648	87396	82030
31327	32037	12666	98858	79028	85930	69897	18842	34078	40364
57192	71834	18796	85698	46440	75389	69006	53679	73873	55622
50971	03014	84836	32146	91046	04530	34510	57698	58800	18404
36183	47640	14699	62828	74408	18609	57774	73178	27402	66123
89886	64379	07806	29447	27901	52683	34391	24074	63355	73392
56652	01510	23253	42137	96275	68326	69326	37652	31579	61152
71113	30478	91123	57893	77707	80281	88562	90784	20593	26343
16287	92673	91972	49899	72351	40716	04840	51805	89032	26051
09532	97306	80316	37964	95312	89290	87715	33847	34525	52460
32571	33947	92181	27933	89863	92116	98107	50498	50066	74269
70723	28350	21723	59834	77513	04946	79159	93619	49999	52596
00798	59561	02754	48896	70545	29415	42955	96808	35539	75565
70554	37739	74633	16604	16957	78825	12281	97451	58000	17959
21221	22001	65815	61172	26374	89651	60355	34946	93525	50499
47938	83086	80065	12685	60966	88424	15563	14883	09418	32657
78793	20115	25696	62515	16022	69013	85950	84771	80589	31355
11211	03557	74341	47499	83763	41973	76609	65934	27460	36842
17383	40211	22636	45264	12674	02747	49033	88406	12135	35578
97519	03601	63457	06591	28801	40508	65772	92916	05157	04066
16215	25044	40916	98535	80918	90021	46173	57901	81776	46643
85416	30889	12546	22118	94502	65908	84198	13151	34073	60104
00060	31408	92929	30875	00163	67257	05808	89117	86191	38445
42077	12050	77916	03126	70043	67694	62817	48086	60158	89377
73381	47171	50607	86418	66657	57161	90907	14141	36777	12228
20621	63455	77117	77509	53256	13163	13577	89998	64450	55918
69671	40340	55271	33239	57703	82068	11146	86896	40107	49964
51133	07432	08421	00072	27981	82383	13822	41872	52249	15662
20056	38288	05522	73457	69982	66421	95874	50316	57143	56466
64162	42447	07134	93865	49453	94743	76272	94179	94144	15476
18840	42299	43880	77485	01657	85980	42637	48175	85231	12539
99160	17627	78709	64998	08790	56697	34231	74398	45608	11193
53122	48168	88234	16684	44204	08816	07391	23138	35905	49207
63164	65958	65135	05121	98591	61629	76757	78506	37450	16262
06540	99772	92779	77541	98300	23034	23988	60257	47229	61954
14912	34023	01413	92760	28289	75153	66918	87284	63366	78284
82005	16101	21534	78440	04273	40139	25783	72745	88375	06806
31890	16803	50250	62500	16677	11571	43340	24022	04759	17658
64933	31340	74968	11913	74572	91949	06025	51238	24986	50864
69020	53846	21913	05884	77506	32118	63703	88960	95514	50487
19012	00300	38428	95279	14910	04514	93069	56734	37129	18880
25328	99686	22973	27889	38340	54954	23783	78794	69366	39116
43438	94254	83791	04992	84608	84781	68302	81709	06031	77354
20873	24753	46084	45133	14972	29953	76644	41496	65481	57039
23977	98084	12333	59677	37748	66490	08490	82283	48881	72748
66183	91635	65518	03702	05771	72251	23296	18735	13995	00753
12313	09536	29445	78796	95320	57752	41853	58244	38246	45413
71806	52188	94321	56546	26579	07219	81925	63367	09107	84799
35977	78248	22956	75073	33979	78146	97786	98384	18699	26512

simsoc

simulated society

2 BLANK FORMS
OTHER MATERIALS

Coordinator's Summary of Operating Information   Form P

1. Size Level

Number of Participants	Size Level
15-32	One
33-47	Two
48-60	Three

3. Munchie Tickets per Session
(SL = Standard of Living)

Size Level	Tickets
One	.7 SL
Two	1.0 SL
Three	1.3 SL

2. Income Consequences if Lowest of National Indicators Is:

Between	Multiplier Is:
80-89	90%
70-79	80%
60-69	70%
50-59	60%
40-49	50%
30-39	40%
20-29	30%
10-19	20%
0-9	10%

Below Zero = Collapse

Between 90-119, basic income remains the same.

When all indicators are above 120, all groups receive 20% more than basic income.

4. Effect of Riots on Social Cohesion (SC)

Each guard post lowers SC by –5.

% of population rioting	5%	10%	15%	20%	25%	30% or more
Effect on SC	0	–2	–6	–12	–20	–30

5. Starting Income for Basic Groups

Group	Size Level One	Two	Three
BASIN, RETSIN	$10	$15	$20
POP, SOP	40	60	80
EMPIN, MASMED, JUDCO	30	45	60

6. BASIN, RETSIN Summary

	One	Two	Three
Starting assets: BASIN	$100	$150	$200
RETSIN	100	150	200
Cost (per passage): BASIN	40	60	80
(per anagram): RETSIN	40	60	80
Payment (per extraction): BASIN	50	75	100
(per marketable word): RETSIN	12	18	24
Maximum for RETSIN	60	90	120
Error deduction (per error): BASIN	4	6	8

BASIN receives no payment if there are six or more errors.

Coordinator's Summary of Operating Information   Form P

1. Size Level

Number of Participants	Size Level
15-32	One
33-47	Two
48-60	Three

2. Income Consequences if Lowest of National Indicators Is:

Between	Multiplier Is:
80-89	90%
70-79	80%
60-69	70%
50-59	60%
40-49	50%
30-39	40%
20-29	30%
10-19	20%
0-9	10%

Below Zero = Collapse

Between 90-119, basic income remains the same.

When all indicators are above 120, all groups receive 20% more than basic income.

3. Munchie Tickets per Session (SL = Standard of Living)

Size Level	Tickets
One	.7 SL
Two	1.0 SL
Three	1.3 SL

4. Effect of Riots on Social Cohesion (SC)

Each guard post lowers SC by -5.

% of population rioting	5%	10%	15%	20%	25%	30% or more
Effect on SC	0	-2	-6	-12	-20	-30

5. Starting Income for Basic Groups

Group	Size Level		
	One	Two	Three
BASIN, RETSIN	$10	$15	$20
POP, SOP	40	60	80
EMPIN, MASMED, JUDCO	30	45	60

6. BASIN, RETSIN Summary

	One	Two	Three
Starting assets: BASIN	$100	$150	$200
RETSIN	100	150	200
Cost (per passage): BASIN	40	60	80
(per anagram): RETSIN	40	60	80
Payment (per extraction): BASIN	50	75	100
(per marketable word): RETSIN	12	18	24
Maximum for RETSIN	60	90	120
Error deduction (per error): BASIN	4	6	8

BASIN receives no payment if there are six or more errors.

Coordinator's Summary of Operating Information Form P

1. Size Level

Number of Participants	Size Level
15-32	One
33-47	Two
48-60	Three

3. Munchie Tickets per Session
(SL = Standard of Living)

Size Level	Tickets
One	.7 SL
Two	1.0 SL
Three	1.3 SL

2. Income Consequences if Lowest of National Indicators Is:

Between	Multiplier Is:
80-89	90%
70-79	80%
60-69	70%
50-59	60%
40-49	50%
30-39	40%
20-29	30%
10-19	20%
0-9	10%

Below Zero = Collapse

Between 90-119, basic income remains the same.

When all indicators are above 120, all groups receive 20% more than basic income.

4. Effect of Riots on Social Cohesion (SC)

Each guard post lowers SC by –5.

% of population rioting	5%	10%	15%	20%	25%	30% or more
Effect on SC	0	–2	–6	–12	–20	–30

5. Starting Income for Basic Groups

Group	Size Level		
	One	Two	Three
BASIN, RETSIN	$10	$15	$20
POP, SOP	40	60	80
EMPIN, MASMED, JUDCO	30	45	60

6. BASIN, RETSIN Summary

	One	Two	Three
Starting assets: BASIN	$100	$150	$200
RETSIN	100	150	200
Cost (per passage): BASIN	40	60	80
(per anagram): RETSIN	40	60	80
Payment (per extraction): BASIN	50	75	100
(per marketable word): RETSIN	12	18	24
Maximum for RETSIN	60	90	120
Error deduction (per error): BASIN	4	6	8

BASIN receives no payment if there are six or more errors.

Coordinator's Summary of Operating Information   Form P

1. Size Level

Number of Participants	Size Level
15-32	One
33-47	Two
48-60	Three

2. Income Consequences if Lowest of National Indicators Is:

Between	Multiplier Is:
80-89	90%
70-79	80%
60-69	70%
50-59	60%
40-49	50%
30-39	40%
20-29	30%
10-19	20%
0-9	10%

Below Zero = Collapse

Between 90-119, basic income remains the same.

When all indicators are above 120, all groups receive 20% more than basic income.

3. Munchie Tickets per Session
(SL = Standard of Living)

Size Level	Tickets
One	.7 SL
Two	1.0 SL
Three	1.3 SL

4. Effect of Riots on Social Cohesion (SC)

Each guard post lowers SC by –5.

% of population rioting	5%	10%	15%	20%	25%	30% or more
Effect on SC	0	–2	–6	–12	–20	–30

5. Starting Income for Basic Groups

Group	One	Two	Three
	Size Level		
BASIN, RETSIN	$10	$15	$20
POP, SOP	40	60	80
EMPIN, MASMED, JUDCO	30	45	60

6. BASIN, RETSIN Summary

	One	Two	Three
Starting assets: BASIN	$100	$150	$200
RETSIN	100	150	200
Cost (per passage): BASIN	40	60	80
(per anagram): RETSIN	40	60	80
Payment (per extraction): BASIN	50	75	100
(per marketable word): RETSIN	12	18	24
Maximum for RETSIN	60	90	120
Error deduction (per error): BASIN	4	6	8

BASIN receives no payment if there are six or more errors.

SIMSOC Roster

Form Q

Num-ber	Name	Region	MGM	Certificates and Agencies	Basic Groups								Session #									
					B	R	P	S	E	M	J	1	2	3	4	5	6	7	8	9	10

KEY

Region

R = Red
Y = Yellow
G = Green
B = Blue

MGM

Check if person is a minority-group member.

Agencies/Certificates

S = Subsistence
T = Travel
L = Luxury Living Endowment
P = Private transportation

Basic Group

B = BASIN
R = RETSIN
P = POP
S = SOP
E = EMPIN
M = MASMED
J = JUDCO

✓ = Employee
H = Head

Subsistence

S = Provided
X = None provided
A = Absent
D = Dead
J = Jailed (under arrest)

SIMSOC Roster

Form Q

Num-ber	Name	Region	MGM	Certificates and Agencies	Basic Groups								Session #									
					B	R	P	S	E	M	J	1	2	3	4	5	6	7	8	9	10

KEY

Region

R = Red
Y = Yellow
G = Green
B = Blue

MGM

Check if person is a minority-group member.

Agencies/Certificates

S = Subsistence
T = Travel
L = Luxury Living Endowment
P = Private transportation

Basic Group

B = BASIN
R = RETSIN
P = POP
S = SOP
E = EMPIN
M = MASMED
J = JUDCO

✓ = Employee
H = Head

Subsistence

S = Provided
X = None provided
A = Absent
D = Dead
J = Jailed (under arrest)

SIMSOC Roster

Form Q

Num-ber	Name	Region	MGM	Certificates and Agencies	Basic Groups								Session #									
					B	R	P	S	E	M	J		1	2	3	4	5	6	7	8	9	10

KEY

Region
R = Red
Y = Yellow
G = Green
B = Blue

MGM
Check if person is a minority-group member.

Agencies/Certificates
S = Subsistence
T = Travel
L = Luxury Living Endowment
P = Private transportation

Basic Group
B = BASIN
R = RETSIN
P = POP
S = SOP
E = EMPIN
M = MASMED
J = JUDCO

✓ = Employee
H = Head

Subsistence
S = Provided
X = None provided
A = Absent
D = Dead
J = Jailed (under arrest)

SIMSOC Roster

Num–ber	Name	Region	MGM	Certificates and Agencies	Basic Groups								Session #									
					B	R	P	S	E	M	J		1	2	3	4	5	6	7	8	9	10

KEY

Region

R = Red
Y = Yellow
G = Green
B = Blue

MGM

Check if person is a minority-group member.

Agencies/Certificates

S = Subsistence
T = Travel
L = Luxury Living Endowment
P = Private transportation

Basic Group

B = BASIN
R = RETSIN
P = POP
S = SOP
E = EMPIN
M = MASMED
J = JUDCO

✓ = Employee
H = Head

Subsistence

S = Provided
X = None provided
A = Absent
D = Dead
J = Jailed (under arrest)

Key for BASIN Passages

Passage No.	A	E	I	O	U
1	22	43	13	17	5
2	13	17	14	4	2
3	14	42	16	17	6
4	18	39	21	29	6
5	29	35	13	21	8
6	18	33	15	17	13
7	37	49	25	25	5
8	31	48	32	43	6
9	27	42	17	24	9
10	24	45	25	22	10
11	24	48	29	29	6
12	23	45	26	29	7
13	23	28	26	28	8
14	34	51	32	33	12
15	22	29	21	17	3
16	32	42	36	34	15
17	28	32	23	21	11
18	32	56	36	25	8
19	32	38	20	21	5
20	26	46	26	26	4
21	18	36	15	22	7
22	25	38	23	22	11
23	25	23	17	21	13
24	23	33	22	21	12
25	17	35	21	31	13
26	19	32	11	22	6
27	28	39	15	19	3
28	15	25	14	14	6
29	30	46	16	29	8
30	12	31	22	23	2
31	18	31	16	19	7
32	31	49	29	42	14
33	31	42	31	26	4
34	23	38	23	21	11
35	13	29	19	19	4
36	21	50	20	23	7
37	19	31	16	24	5
38	24	37	19	21	7
39	26	38	8	16	9
40	15	23	24	17	6

RETSIN Anagrams and Marketable Words   Form U

1.	MEPCALN	camel, clean, pecan, ample, place
2.	ULPEGAR	argue, purge, grape, pearl, glare
3.	ENTORS	store, onset, tenor, stone, snore
4.	COPATIN	panic, paint, antic, topic, tonic
5.	ASETEL	lease, slate, least, tease, easel
6.	ENVERS	serve, sever, verse, nerve, sneer
7.	THEWBIR	birth, their, tribe, write, white
8.	ENKTAS	steak, skate, snake, sneak, taken
9.	ENDIRF	fiend, fined, infer, finer, fired
10.	NERTAG	agent, range, great, grant, anger
11.	ADRENCY	dance, candy, decay, yearn, ready
12.	PETQUIS	equip, quiet, spite, upset, quest
13.	IDETBOR	orbit, tribe, tired, bored, bride
14.	UDCALEN	dunce, uncle, clean, caned, decal
15.	SHETUGO	ghost, shout, ought, house, gusto
16.	SNETHEV	seven, event, sheet, these, tense
17.	MONWELB	elbow, blown, lemon, noble, below
18.	GENLELA	eagle, angel, glean, angle, legal
19.	TASYEL	style, yeast, steal, stale, least
20.	DAYRETH	tardy, ready, earth, heart, hardy
21.	YILALED	alley, daily, delay, yield, ladle
22.	EUROPAS	purse, super, prose, opera, spare
23.	ORSTETA	taste, state, otter, stare, store
24.	RICPOHE	choir, porch, price, chore, perch
25.	NIFLAPT	final, plain, plant, flint, paint
26.	ACPERES	spear, peace, space, spare, creep
27.	REVCABD	brave, crave, beard, bread, brace
28.	TARFONE	often, forte, front, after, tenor
29.	PLUTSIO	tulip, pilot, spoil, spilt, split
30.	NIRGEDA	ridge, range, anger, grain, grade
31.	HEXGITS	sixth, exist, heist, eight, sight
32.	TARWES	water, straw, waste, stare, sweat
33.	KERITCA	trick, track, trace, crate, cater
34.	SUROMPT	tumor, sport, storm, strum, spout
35.	RELPLYA	early, pearl, rally, alley, lapel
36.	KEBSORIT	store, stork, broke, orbit, tribe
37.	NEREFIG	infer, finer, reign, green, grief
38.	TALKES	steak, skate, steal, stale, slate
39.	SELYREV	serve, sever, verse, lever, every
40.	ZURESPIE	purse, super, prize, seize, spire

Investment Tally Sheet

Form V

Session	BASIN Passages						RETSIN Anagrams						Public Programs	
	1	2	3	4	5	Total	1	2	3	4	5	Total	R&C [a]	WS [b]
1.														
2.														
3.														
4.														
5.														
6.														
7.														
8.														
9.														
10.														

Payment Schedule

	Size Level		
	One	Two	Three
BASIN (per passage)	50	75	100
BASIN error deduction (per error)	4	6	8
RETSIN (per marketable word)	12	18	24

a R&C = Research and Conservation
b WS = Welfare Services

National Indicator Calculations

Form W

Session # _____

National Indicators

	FES	SL	SC	PC
1. Initial value (at beginning of session)	____	____	____	____
2. Natural decline (10% of line 1)	− ____	− ____	− ____	− ____
3. Research and Conservation = _____ (FES = +40%; SL = +10%)	+ ____	+ ____	0	0
4. Welfare Services = _____ (SL = +10%; SC, PC = +20%)	0	+ ____	+ ____	+ ____
5. BASIN passages bought = _____ (FES = −2; SL = +1)	− ____	+ ____	0	0
6. RETSIN anagrams bought = _____ (SL = +1; PC = −1)	0	+ ____	0	− ____
7. Number of absentees (Ab) = _____ (SL, PC = −2Ab)	0	− ____	0	− ____
8. Number of unemployed (U) = _____ (SL, SC = −3U; PC = −1U)	0	− ____	− ____	− ____
9. Number of rioters (R) = _____ (For SC, see Form P; PC = −2R)	0	0	− ____	− ____
10. Number of arrests (Ar) = _____ (SC, PC = −3Ar)	0	0	− ____	− ____
11. Number of deaths (D) = _____ (SL, SC, PC = −5D)	0	− ____	− ____	− ____
12. Individual goal declarations: Positive (P) = _____; Negative (N) = _____ (PC = .25P−N)	0	0	0	± ____
13. Special events	____	____	____	____
14. Total of all pluses	____	____	____	____
15. Total of all minuses	____	____	____	____
16. Net change (add lines 14 & 15)	____	____	____	____
17. Final value (add lines 1 & 16)	____	____	____	____

National Indicator Calculations

Form W

Session # _____

National Indicators

	FES	SL	SC	PC
1. Initial value (at beginning of session)	____	____	____	____
2. Natural decline (10% of line 1)	−____	−____	−____	−____
3. Research and Conservation = _____ (FES = +40%; SL = +10%)	+____	+____	0	0
4. Welfare Services = _____ (SL = +10%; SC, PC = +20%)	0	+____	+____	+____
5. BASIN passages bought = _____ (FES = −2; SL = +1)	−____	+____	0	0
6. RETSIN anagrams bought = _____ (SL = +1; PC = −1)	0	+____	0	−____
7. Number of absentees (Ab) = _____ (SL, PC = −2Ab)	0	−____	0	−____
8. Number of unemployed (U) = _____ (SL, SC = −3U; PC = −1U)	0	−____	−____	−____
9. Number of rioters (R) = _____ (For SC, see Form P; PC = −2R)	0	0	−____	−____
10. Number of arrests (Ar) = _____ (SC, PC = −3Ar)	0	0	−____	−____
11. Number of deaths (D) = _____ (SL, SC, PC = −5D)	0	−____	−____	−____
12. Individual goal declarations: Positive (P) = _____; Negative (N) = _____ (PC = .25P−N)	0	0	0	±____
13. Special events	____	____	____	____
14. Total of all pluses	____	____	____	____
15. Total of all minuses	____	____	____	____
16. Net change (add lines 14 & 15)	____	____	____	____
17. Final value (add lines 1 & 16)	____	____	____	____

National Indicator Calculations

Form W

Session # _____

National Indicators

	FES	SL	SC	PC
1. Initial value (at beginning of session)	———	———	———	———
2. Natural decline (10% of line 1)	–———	–———	–———	–———
3. Research and Conservation = _____ (FES = +40%; SL = +10%)	+———	+———	0	0
4. Welfare Services = _____ (SL = +10%; SC, PC = +20%)	0	+———	+———	+———
5. BASIN passages bought = _____ (FES = –2; SL = +1)	–———	+———	0	0
6. RETSIN anagrams bought = _____ (SL = +1; PC = –1)	0	+———	0	–
7. Number of absentees (Ab) = _____ (SL, PC = –2Ab)	0	–———	0	–———
8. Number of unemployed (U) = _____ (SL, SC = –3U; PC = –1U)	0	–———	–———	–———
9. Number of rioters (R) = _____ (For SC, see Form P; PC = –2R)	0	0	–———	–———
10. Number of arrests (Ar) = _____ (SC, PC = –3Ar)	0	0	–———	–———
11. Number of deaths (D) = _____ (SL, SC, PC = –5D)	0	–———	–———	–———
12. Individual goal declarations: Positive (P) = _____; Negative (N) = _____ (PC = .25P-N)	0	0	0	+———
13. Special events	———	———	———	———
14. Total of all pluses	———	———	———	———
15. Total of all minuses	———	———	———	———
16. Net change (add lines 14 & 15)	———	———	———	———
17. Final value (add lines 1 & 16)	———	———	———	———

National Indicator Calculations

Form W

Session # _____

National Indicators

	FES	SL	SC	PC
1. Initial value (at beginning of session)	_____	_____	_____	_____
2. Natural decline (10% of line 1)	− _____	− _____	− _____	− _____
3. Research and Conservation = _____ (FES = +40%; SL = +10%)	+ _____	+ _____	0	0
4. Welfare Services = _____ (SL = +10%; SC, PC = +20%)	0	+ _____	+ _____	+ _____
5. BASIN passages bought = _____ (FES = −2; SL = +1)	− _____	+ _____	0	0
6. RETSIN anagrams bought = _____ (SL = +1; PC = −1)	0	+ _____	0	− _____
7. Number of absentees (Ab) = _____ (SL, PC = −2Ab)	0	− _____	0	− _____
8. Number of unemployed (U) = _____ (SL, SC = −3U; PC = −1U)	0	− _____	− _____	− _____
9. Number of rioters (R) = _____ (For SC, see Form P; PC = −2R)	0	0	− _____	− _____
10. Number of arrests (Ar) = _____ (SC, PC = −3Ar)	0	0	− _____	− _____
11. Number of deaths (D) = _____ (SL, SC, PC = −5D)	0	− _____	− _____	− _____
12. Individual goal declarations: Positive (P) = _____; Negative (N) = _____ (PC = .25P−N)	0	0	0	± _____
13. Special events	_____	_____	_____	_____
14. Total of all pluses	_____	_____	_____	_____
15. Total of all minuses	_____	_____	_____	_____
16. Net change (add lines 14 & 15)	_____	_____	_____	_____
17. Final value (add lines 1 & 16)	_____	_____	_____	_____

National Indicator Calculations

Session # _____

National Indicators

	FES	SL	SC	PC
1. Initial value (at beginning of session)	_____	_____	_____	_____
2. Natural decline (10% of line 1)	– _____	– _____	– _____	– _____
3. Research and Conservation = _____ (FES = +40%; SL = +10%)	+ _____	+ _____	0	0
4. Welfare Services = _____ (SL = +10%; SC, PC = +20%)	0	+ _____	+ _____	+ _____
5. BASIN passages bought = _____ (FES = –2; SL = +1)	– _____	+ _____	0	0
6. RETSIN anagrams bought = _____ (SL = +1; PC = –1)	0	+ _____	0	– _____
7. Number of absentees (Ab) = _____ (SL, PC = –2Ab)	0	– _____	0	– _____
8. Number of unemployed (U) = _____ (SL, SC = –3U; PC = –1U)	0	– _____	– _____	– _____
9. Number of rioters (R) = _____ (For SC, see Form P; PC = –2R)	0	0	– _____	– _____
10. Number of arrests (Ar) = _____ (SC, PC = –3Ar)	0	0	– _____	– _____
11. Number of deaths (D) = _____ (SL, SC, PC = –5D)	0	– _____	– _____	– _____
12. Individual goal declarations: Positive (P) = _____; Negative (N) = _____ (PC = .25P-N)	0	0	0	+ _____ –
13. Special events	_____	_____	_____	_____
14. Total of all pluses	_____	_____	_____	_____
15. Total of all minuses	_____	_____	_____	_____
16. Net change (add lines 14 & 15)	_____	_____	_____	_____
17. Final value (add lines 1 & 16)	_____	_____	_____	_____

National Indicator Calculations

Form W

Session # _____

National Indicators

		FES	SL	SC	PC
1.	Initial value (at beginning of session)	_____	_____	_____	_____
2.	Natural decline (10% of line 1)	– _____	– _____	– _____	– _____
3.	Research and Conservation = _____ (FES = +40%; SL = +10%)	+ _____	+ _____	0 _____	0 _____
4.	Welfare Services = _____ (SL = +10%; SC, PC = +20%)	0 _____	+ _____	+ _____	+ _____
5.	BASIN passages bought = _____ (FES = –2; SL = +1)	– _____	+ _____	0 _____	0 _____
6.	RETSIN anagrams bought = _____ (SL = +1; PC = –1)	0 _____	+ _____	0 _____	– _____
7.	Number of absentees (Ab) = _____ (SL, PC = –2Ab)	0 _____	– _____	0 _____	– _____
8.	Number of unemployed (U) = _____ (SL, SC = –3U; PC = –1U)	0 _____	– _____	– _____	– _____
9.	Number of rioters (R) = _____ (For SC, see Form P; PC = –2R)	0 _____	0 _____	– _____	– _____
10.	Number of arrests (Ar) = _____ (SC, PC = –3Ar)	0 _____	0 _____	– _____	– _____
11.	Number of deaths (D) = _____ (SL, SC, PC = –5D)	0 _____	– _____	– _____	– _____
12.	Individual goal declarations: Positive (P) = _____; Negative (N) = _____ (PC = .25P-N)	0 _____	0 _____	0 _____	$\overset{+}{-}$ _____
13.	Special events	_____	_____	_____	_____
14.	Total of all pluses	_____	_____	_____	_____
15.	Total of all minuses	_____	_____	_____	_____
16.	Net change (add lines 14 & 15)	_____	_____	_____	_____
17.	Final value (add lines 1 & 16)	_____	_____	_____	_____

National Indicator Calculations

Form W

Session # _____

National Indicators

	FES	SL	SC	PC
1. Initial value (at beginning of session)	_____	_____	_____	_____
2. Natural decline (10% of line 1)	– _____	– _____	– _____	– _____
3. Research and Conservation = _____ (FES = +40%; SL = +10%)	+ _____	+ _____	0 _____	0 _____
4. Welfare Services = _____ (SL = +10%; SC, PC = +20%)	0 _____	+ _____	+ _____	+ _____
5. BASIN passages bought = _____ (FES = –2; SL = +1)	– _____	+ _____	0 _____	0 _____
6. RETSIN anagrams bought = _____ (SL = +1; PC = –1)	0 _____	+ _____	0 _____	– _____
7. Number of absentees (Ab) = _____ (SL, PC = –2Ab)	0 _____	– _____	0 _____	– _____
8. Number of unemployed (U) = _____ (SL, SC = –3U; PC = –1U)	0 _____	– _____	– _____	– _____
9. Number of rioters (R) = _____ (For SC, see Form P; PC = –2R)	0 _____	0 _____	– _____	– _____
10. Number of arrests (Ar) = _____ (SC, PC = –3Ar)	0 _____	0 _____	– _____	– _____
11. Number of deaths (D) = _____ (SL, SC, PC = –5D)	0 _____	– _____	– _____	– _____
12. Individual goal declarations: Positive (P) = _____; Negative (N) = _____ (PC = .25P-N)	0 _____	0 _____	0 _____	± _____
13. Special events	_____	_____	_____	_____
14. Total of all pluses	_____	_____	_____	_____
15. Total of all minuses	_____	_____	_____	_____
16. Net change (add lines 14 & 15)	_____	_____	_____	_____
17. Final value (add lines 1 & 16)	_____	_____	_____	_____

National Indicator Calculations

Form W

Session # _____

National Indicators

	FES	SL	SC	PC
1. Initial value (at beginning of session)	_____	_____	_____	_____
2. Natural decline (10% of line 1)	–_____	–_____	–_____	–_____
3. Research and Conservation = _____ (FES = +40%; SL = +10%)	+_____	+_____	0	0
4. Welfare Services = _____ (SL = +10%; SC, PC = +20%)	0	+_____	+_____	+_____
5. BASIN passages bought = _____ (FES = –2; SL = +1)	–_____	+_____	0	0
6. RETSIN anagrams bought = _____ (SL = +1; PC = –1)	0	+_____	0	–_____
7. Number of absentees (Ab) = _____ (SL, PC = –2Ab)	0	–_____	0	–_____
8. Number of unemployed (U) = _____ (SL, SC = –3U; PC = –1U)	0	–_____	–_____	–_____
9. Number of rioters (R) = _____ (For SC, see Form P; PC = –2R)	0	0	–_____	–_____
10. Number of arrests (Ar) = _____ (SC, PC = –3Ar)	0	0	–_____	–_____
11. Number of deaths (D) = _____ (SL, SC, PC = –5D)	0	–_____	–_____	–_____
12. Individual goal declarations: Positive (P) = _____; Negative (N) = _____ (PC = .25P-N)	0	0	0	+_____ –
13. Special events	_____	_____	_____	_____
14. Total of all pluses	_____	_____	_____	_____
15. Total of all minuses	_____	_____	_____	_____
16. Net change (add lines 14 & 15)	_____	_____	_____	_____
17. Final value (add lines 1 & 16)	_____	_____	_____	_____

National Indicator Calculations

Form W

Session # _____

National Indicators

	FES	SL	SC	PC
1. Initial value (at beginning of session)	____	____	____	____
2. Natural decline (10% of line 1)	− ____	− ____	− ____	− ____
3. Research and Conservation = _____ (FES = +40%; SL = +10%)	+ ____	+ ____	0	0
4. Welfare Services = _____ (SL = +10%; SC, PC = +20%)	0	+ ____	+ ____	+ ____
5. BASIN passages bought = _____ (FES = −2; SL = +1)	− ____	+ ____	0	0
6. RETSIN anagrams bought = _____ (SL = +1; PC = −1)	0	+ ____	0	− ____
7. Number of absentees (Ab) = _____ (SL, PC = −2Ab)	0	− ____	0	− ____
8. Number of unemployed (U) = _____ (SL, SC = −3U; PC = −1U)	0	− ____	− ____	− ____
9. Number of rioters (R) = _____ (For SC, see Form P; PC = −2R)	0	0	− ____	− ____
10. Number of arrests (Ar) = _____ (SC, PC = −3Ar)	0	0	− ____	− ____
11. Number of deaths (D) = _____ (SL, SC, PC = −5D)	0	− ____	− ____	− ____
12. Individual goal declarations: Positive (P) = _____; Negative (N) = _____ (PC = .25P−N)	0	0	0	± ____
13. Special events	____	____	____	____
14. Total of all pluses	____	____	____	____
15. Total of all minuses	____	____	____	____
16. Net change (add lines 14 & 15)	____	____	____	____
17. Final value (add lines 1 & 16)	____	____	____	____

National Indicator Calculations

Form W

Session # _____

National Indicators

	FES	SL	SC	PC
1. Initial value (at beginning of session)	_____	_____	_____	_____
2. Natural decline (10% of line 1)	- _____	- _____	- _____	- _____
3. Research and Conservation = _____ (FES = +40%; SL = +10%)	+ _____	+ _____	0	0
4. Welfare Services = _____ (SL = +10%; SC, PC = +20%)	0	+ _____	+ _____	+ _____
5. BASIN passages bought = _____ (FES = -2; SL = +1)	- _____	+ _____	0	0
6. RETSIN anagrams bought = _____ (SL = +1; PC = -1)	0	+ _____	0	- _____
7. Number of absentees (Ab) = _____ (SL, PC = -2Ab)	0	- _____	0	- _____
8. Number of unemployed (U) = _____ (SL, SC = -3U; PC = -1U)	0	- _____	- _____	- _____
9. Number of rioters (R) = _____ (For SC, see Form P; PC = -2R)	0	0	- _____	- _____
10. Number of arrests (Ar) = _____ (SC, PC = -3Ar)	0	0	- _____	- _____
11. Number of deaths (D) = _____ (SL, SC, PC = -5D)	0	- _____	- _____	- _____
12. Individual goal declarations: Positive (P) = _____; Negative (N) = _____ (PC = .25P-N)	0	0	0	± _____
13. Special events	_____	_____	_____	_____
14. Total of all pluses	_____	_____	_____	_____
15. Total of all minuses	_____	_____	_____	_____
16. Net change (add lines 14 & 15)	_____	_____	_____	_____
17. Final value (add lines 1 & 16)	_____	_____	_____	_____

National Indicator Calculations

Form W

Session # _____

National Indicators

	FES	SL	SC	PC
1. Initial value (at beginning of session)	____	____	____	____
2. Natural decline (10% of line 1)	____	____	____	____
3. Research and Conservation = _____ (FES = +40%; SL = +10%)	+	+	0	0
4. Welfare Services = _____ (SL = +10%; SC, PC = +20%)	0	+	+	+
5. BASIN passages bought = _____ (FES = –2; SL = +1)	–	+	0	0
6. RETSIN anagrams bought = _____ (SL = +1; PC = –1)	0	+	0	–
7. Number of absentees (Ab) = _____ (SL, PC = –2Ab)	0	–	0	–
8. Number of unemployed (U) = _____ (SL, SC = –3U; PC = –1U)	0	–	–	–
9. Number of rioters (R) = _____ (For SC, see Form P; PC = –2R)	0	0	–	–
10. Number of arrests (Ar) = _____ (SC, PC = –3Ar)	0	0	–	–
11. Number of deaths (D) = _____ (SL, SC, PC = –5D)	0	–	–	–
12. Individual goal declarations: Positive (P) = _____; Negative (N) = _____ (PC = .25P-N)	0	0	0	$\overset{+}{-}$
13. Special events	____	____	____	____
14. Total of all pluses	____	____	____	____
15. Total of all minuses	____	____	____	____
16. Net change (add lines 14 & 15)	____	____	____	____
17. Final value (add lines 1 & 16)	____	____	____	____

National Indicator Calculations

Form W

Session # _____

National Indicators

		FES	SL	SC	PC
1.	Initial value (at beginning of session)	_____	_____	_____	_____
2.	Natural decline (10% of line 1)	− _____	− _____	− _____	− _____
3.	Research and Conservation = _____ (FES = +40%; SL = +10%)	+ _____	+ _____	0	0
4.	Welfare Services = _____ (SL = +10%; SC, PC = +20%)	0	+ _____	+ _____	+ _____
5.	BASIN passages bought = _____ (FES = −2; SL = +1)	− _____	+ _____	0	0
6.	RETSIN anagrams bought = _____ (SL = +1; PC = −1)	0	+ _____	0	− _____
7.	Number of absentees (Ab) = _____ (SL, PC = −2Ab)	0	− _____	0	− _____
8.	Number of unemployed (U) = _____ (SL, SC = −3U; PC = −1U)	0	− _____	− _____	− _____
9.	Number of rioters (R) = _____ (For SC, see Form P; PC = −2R)	0	0	− _____	− _____
10.	Number of arrests (Ar) = _____ (SC, PC = −3Ar)	0	0	− _____	− _____
11.	Number of deaths (D) = _____ (SL, SC, PC = −5D)	0	− _____	− _____	− _____
12.	Individual goal declarations: Positive (P) = _____; Negative (N) = _____ (PC = .25P−N)	0	0	0	± _____
13.	Special events	_____	_____	_____	_____
14.	Total of all pluses	_____	_____	_____	_____
15.	Total of all minuses	_____	_____	_____	_____
16.	Net change (add lines 14 & 15)	_____	_____	_____	_____
17.	Final value (add lines 1 & 16)	_____	_____	_____	_____

National Indicator Calculations

Session # _____

National Indicators

	FES	SL	SC	PC
1. Initial value (at beginning of session)	_____	_____	_____	_____
2. Natural decline (10% of line 1)	– _____	– _____	– _____	– _____
3. Research and Conservation = _____ (FES = +40%; SL = +10%)	+ _____	+ _____	0	0
4. Welfare Services = _____ (SL = +10%; SC, PC = +20%)	0	+ _____	+ _____	+ _____
5. BASIN passages bought = _____ (FES = –2; SL = +1)	– _____	+ _____	0	0
6. RETSIN anagrams bought = _____ (SL = +1; PC = –1)	0	+ _____	0	– _____
7. Number of absentees (Ab) = _____ (SL, PC = –2Ab)	0	– _____	0	– _____
8. Number of unemployed (U) = _____ (SL, SC = –3U; PC = –1U)	0	– _____	– _____	– _____
9. Number of rioters (R) = _____ (For SC, see Form P; PC = –2R)	0	0	– _____	– _____
10. Number of arrests (Ar) = _____ (SC, PC = –3Ar)	0	0	– _____	– _____
11. Number of deaths (D) = _____ (SL, SC, PC = –5D)	0	– _____	– _____	– _____
12. Individual goal declarations: Positive (P) = _____; Negative (N) = _____ (PC = .25P-N)	0	0	0	± _____
13. Special events	_____	_____	_____	_____
14. Total of all pluses	_____	_____	_____	_____
15. Total of all minuses	_____	_____	_____	_____
16. Net change (add lines 14 & 15)	_____	_____	_____	_____
17. Final value (add lines 1 & 16)	_____	_____	_____	_____

Basic Group Income Calculations

Form X

Session # _____

	BASIN	RETSIN	POP	SOP	EMPIN	MASMED	JUDCO
1. Assets at beginning of session	——	——					
2. Assets withdrawn on Form I	——	——					
3. Payments due for product (Totals from Form V)	——	——					
4. Net assets in bank	——	——					
5. Support cards turned in			——	——	——	——	
6. Starting income [a]			——	——			
7. Support product (multiply line 5 by line 6)			——	——			
8. Total population (TP) = _____ (divide line 7 by TP)			——	——			
9. Support multiplier			2.5	2.5	2	2	
10. Basic income [b]	——	——	——	——	——	——	——
11. National Indicator multiplier [c] (_____%)							
12. Net Income (line 10 × line 11)	——	——	——	——	——	——	——

[a] For POP & SOP: Size Level One = $40
                             Two = $60
                           Three = $80

[b] Basic Income:
  For BASIN/RETSIN = 10% of line 4
  For POP/SOP   = line 8 × line 9
  For EMPIN/MASMED = line 5 × line 9
  For JUDCO   = Size Level One = $30
                               Two = $45
                               Three = $60

[c] Use new National Indicator levels from Form W and Table 2 from Form P to obtain correct figure.

Basic Group Income Calculations

Form X

Session # _____

	BASIN	RETSIN	POP	SOP	EMPIN	MASMED	JUDCO
1. Assets at beginning of session	——	——					
2. Assets withdrawn on Form I	——	——					
3. Payments due for product (Totals from Form V)	——	——					
4. Net assets in bank	——	——					
5. Support cards turned in			——	——	——	——	
6. Starting income [a]			——	——			
7. Support product (multiply line 5 by line 6)			——	——			
8. Total population (TP) = _____ (divide line 7 by TP)			——	——			
9. Support multiplier			2.5	2.5	2	2	
10. Basic income [b]	——	——	——	——	——	——	——
11. National Indicator multiplier [c] (_____%)							
12. Net Income (line 10 × line 11)	——	——	——	——	——	——	——

[a] For POP & SOP: Size Level One = $40
                           Two = $60
                           Three = $80

[b] Basic Income:
   For BASIN/RETSIN   = 10% of line 4
   For POP/SOP   = line 8 × line 9
   For EMPIN/MASMED   = line 5 × line 9
   For JUDCO   = Size Level One = $30
                                   Two = $45
                                   Three = $60

[c] Use new National Indicator levels from Form W and Table 2 from Form P to obtain correct figure.

Basic Group Income Calculations

Form X

Session # _____

	BASIN	RETSIN	POP	SOP	EMPIN	MASMED	JUDCO
1. Assets at beginning of session	—	—					
2. Assets withdrawn on Form I	—	—					
3. Payments due for product (Totals from Form V)	—	—					
4. Net assets in bank	—	—					
5. Support cards turned in			—	—	—	—	
6. Starting income [a]			—	—			
7. Support product (multiply line 5 by line 6)			—	—			
8. Total population (TP) = _____ (divide line 7 by TP)			—	—			
9. Support multiplier			2.5	2.5	2	2	
10. Basic income [b]	—	—	—	—	—	—	—
11. National Indicator multiplier [c] (_____%)							
12. Net Income (line 10 × line 11)	—	—	—	—	—	—	—

[a] For POP & SOP: Size Level One = $40
             Two = $60
             Three = $80

[b] Basic Income:
    For BASIN/RETSIN   = 10% of line 4
    For POP/SOP   = line 8 × line 9
    For EMPIN/MASMED   = line 5 × line 9
    For JUDCO   = Size Level One = $30
                                    Two = $45
                                    Three = $60

[c] Use new National Indicator levels from Form W and Table 2 from Form P to obtain correct figure.

Basic Group Income Calculations

Form X

	BASIN	RETSIN	POP	SOP	EMPIN	MASMED	JUDCO
1. Assets at beginning of session	—	—					
2. Assets withdrawn on Form I	—	—					
3. Payments due for product (Totals from Form V)	—	—					
4. Net assets in bank	—	—					
5. Support cards turned in			—	—	—	—	
6. Starting income [a]			—	—			
7. Support product (multiply line 5 by line 6)			—	—			
8. Total population (TP) = _____ (divide line 7 by TP)			—	—			
9. Support multiplier			2.5	2.5	2	2	
10. Basic income [b]	—	—	—	—	—	—	—
11. National Indicator multiplier [c] (____%)							
12. Net Income (line 10 × line 11)	—	—	—	—	—	—	—

[a] For POP & SOP: Size Level One = $40
Two = $60
Three = $80

[b] Basic Income:
For BASIN/RETSIN = 10% of line 4
For POP/SOP = line 8 × line 9
For EMPIN/MASMED = line 5 × line 9
For JUDCO = Size Level One = $30
Two = $45
Three = $60

[c] Use new National Indicator levels from Form W and Table 2 from Form P to obtain correct figure.

Basic Group Income Calculations

Form X

Session # _____

	BASIN	RETSIN	POP	SOP	EMPIN	MASMED	JUDCO
1. Assets at beginning of session	——	——					
2. Assets withdrawn on Form I	——	——					
3. Payments due for product (Totals from Form V)	——	——					
4. Net assets in bank	——	——					
5. Support cards turned in			——	——	——	——	
6. Starting income [a]			——	——			
7. Support product (multiply line 5 by line 6)			——	——			
8. Total population (TP) = _____ (divide line 7 by TP)			——	——			
9. Support multiplier			2.5	2.5	2	2	
10. Basic income [b]	——	——	——	——	——	——	——
11. National Indicator multiplier [c] (____%)							
12. Net Income (line 10 × line 11)	——	——	——	——	——	——	——

[a] For POP & SOP: Size Level One = $40
                              Two = $60
                              Three = $80

[b] Basic Income:
   For BASIN/RETSIN   = 10% of line 4
   For POP/SOP   = line 8 × line 9
   For EMPIN/MASMED = line 5 × line 9
   For JUDCO   = Size Level One = $30
                                        Two = $45
                                        Three = $60

[c] Use new National Indicator levels from Form W and Table 2 from Form P to obtain correct figure.

Basic Group Income Calculations

Form X

Session # _____

	BASIN	RETSIN	POP	SOP	EMPIN	MASMED	JUDCO
1. Assets at beginning of session	——	——					
2. Assets withdrawn on Form I	——	——					
3. Payments due for product (Totals from Form V)	——	——					
4. Net assets in bank	——	——					
5. Support cards turned in			——	——	——	——	
6. Starting income [a]			——	——			
7. Support product (multiply line 5 by line 6)			——	——			
8. Total population (TP) = _____ (divide line 7 by TP)			——	——			
9. Support multiplier			2.5	2.5	2	2	
10. Basic income [b]	——	——	——	——	——	——	——
11. National Indicator multiplier [c] (____%)							
12. Net Income (line 10 × line 11)	——	——	——	——	——	——	——

[a] For POP & SOP: Size Level One = $40
　　　　　　　　　　Two = $60
　　　　　　　　　　Three = $80

[b] Basic Income:
　For BASIN/RETSIN　= 10% of line 4
　For POP/SOP　　　= line 8 × line 9
　For EMPIN/MASMED = line 5 × line 9
　For JUDCO　　　　= Size Level One = $30
　　　　　　　　　　　　　　Two = $45
　　　　　　　　　　　　　　Three = $60

[c] Use new National Indicator levels from Form W and Table 2 from Form P to obtain correct figure.

Basic Group Income Calculations

Form X

Session # _____

	BASIN	RETSIN	POP	SOP	EMPIN	MASMED	JUDCO
1. Assets at beginning of session	——	——					
2. Assets withdrawn on Form I	——	——					
3. Payments due for product (Totals from Form V)	——	——					
4. Net assets in bank	——	——					
5. Support cards turned in			——	——	——	——	
6. Starting income [a]			——	——			
7. Support product (multiply line 5 by line 6)			——	——			
8. Total population (TP) = _____ (divide line 7 by TP)			——	——			
9. Support multiplier			2.5	2.5	2	2	
10. Basic income [b]	——	——	——	——	——	——	——
11. National Indicator multiplier [c] (____%)							
12. Net Income (line 10 × line 11)	——	——	——	——	——	——	——

[a] For POP & SOP: Size Level One = $40
  Two = $60
  Three = $80

[b] Basic Income:
  For BASIN/RETSIN = 10% of line 4
  For POP/SOP = line 8 × line 9
  For EMPIN/MASMED = line 5 × line 9
  For JUDCO = Size Level One = $30
    Two = $45
    Three = $60

[c] Use new National Indicator levels from Form W and Table 2 from Form P to obtain correct figure.

Basic Group Income Calculations

Form X

Session # _____

	BASIN	RETSIN	POP	SOP	EMPIN	MASMED	JUDCO
1. Assets at beginning of session	——	——					
2. Assets withdrawn on Form I	——	——					
3. Payments due for product (Totals from Form V)	——	——					
4. Net assets in bank	——	——					
5. Support cards turned in			——	——	——	——	
6. Starting income [a]			——	——			
7. Support product (multiply line 5 by line 6)			——	——			
8. Total population (TP) = _____ (divide line 7 by TP)			——	——			
9. Support multiplier			2.5	2.5	2	2	
10. Basic income [b]	——	——	——	——	——	——	——
11. National Indicator multiplier [c] (_____%)							
12. Net Income (line 10 × line 11)	——	——	——	——	——	——	——

[a] For POP & SOP: Size Level One = $40
       Two = $60
       Three = $80

[b] Basic Income:
 For BASIN/RETSIN = 10% of line 4
 For POP/SOP   = line 8 × line 9
 For EMPIN/MASMED = line 5 × line 9
 For JUDCO   = Size Level One = $30
           Two = $45
           Three = $60

[c] Use new National Indicator levels from Form W and Table 2 from Form P to obtain correct figure.

Basic Group Income Calculations

Form X

Session # _____

	BASIN	RETSIN	POP	SOP	EMPIN	MASMED	JUDCO
1. Assets at beginning of session	——	——					
2. Assets withdrawn on Form I	——	——					
3. Payments due for product (Totals from Form V)	——	——					
4. Net assets in bank	——	——					
5. Support cards turned in			——	——	——	——	
6. Starting income [a]			——	——			
7. Support product (multiply line 5 by line 6)			——	——			
8. Total population (TP) = _____ (divide line 7 by TP)			——	——			
9. Support multiplier			2.5	2.5	2	2	
10. Basic income [b]	——	——	——	——	——	——	——
11. National Indicator multiplier [c] (_____%)							
12. Net Income (line 10 × line 11)	——	——	——	——	——	——	——

[a] For POP & SOP: Size Level One = $40
                            Two = $60
                          Three = $80

[b] Basic Income:
  For BASIN/RETSIN   = 10% of line 4
  For POP/SOP   = line 8 × line 9
  For EMPIN/MASMED = line 5 × line 9
  For JUDCO   = Size Level One = $30
                               Two = $45
                               Three = $60

[c] Use new National Indicator levels from Form W and Table 2 from Form P to obtain correct figure.

Basic Group Income Calculations

	BASIN	RETSIN	POP	SOP	EMPIN	MASMED	JUDCO
1. Assets at beginning of session	——	——					
2. Assets withdrawn on Form I	——	——					
3. Payments due for product (Totals from Form V)	——	——					
4. Net assets in bank	——	——					
5. Support cards turned in			——	——	——	——	
6. Starting income [a]			——	——			
7. Support product (multiply line 5 by line 6)			——	——			
8. Total population (TP) = _____ (divide line 7 by TP)			——	——			
9. Support multiplier			2.5	2.5	2	2	
10. Basic income [b]	——	——	——	——	——	——	——
11. National Indicator multiplier [c] (____%)							
12. Net Income (line 10 × line 11)	——	——	——	——	——	——	——

[a] For POP & SOP: Size Level One = $40
                               Two = $60
                               Three = $80

[b] Basic Income:
  For BASIN/RETSIN   = 10% of line 4
  For POP/SOP   = line 8 × line 9
  For EMPIN/MASMED = line 5 × line 9
  For JUDCO   = Size Level One = $30
                              Two = $45
                              Three = $60

[c] Use new National Indicator levels from Form W and Table 2 from Form P to obtain correct figure.

Basic Group Income Calculations

Form X

Session # _____

	BASIN	RETSIN	POP	SOP	EMPIN	MASMED	JUDCO
1. Assets at beginning of session	___	___					
2. Assets withdrawn on Form I	___	___					
3. Payments due for product (Totals from Form V)	___	___					
4. Net assets in bank	___	___					
5. Support cards turned in			___	___	___	___	
6. Starting income [a]			___	___			
7. Support product (multiply line 5 by line 6)			___	___			
8. Total population (TP) = _____ (divide line 7 by TP)			___	___			
9. Support multiplier			2.5	2.5	2	2	
10. Basic income [b]	___	___	___	___	___	___	___
11. National Indicator multiplier [c] (_____%)	___	___	___	___	___	___	___
12. Net Income (line 10 × line 11)	___	___	___	___	___	___	___

[a] For POP & SOP: Size Level One = $40
                     Two = $60
                     Three = $80

[b] Basic Income:
  For BASIN/RETSIN   = 10% of line 4
  For POP/SOP   = line 8 × line 9
  For EMPIN/MASMED = line 5 × line 9
  For JUDCO   = Size Level One = $30
                             Two = $45
                             Three = $60

[c] Use new National Indicator levels from Form W and Table 2 from Form P to obtain correct figure.

Basic Group Income Calculations

Form X

Session # _____

	BASIN	RETSIN	POP	SOP	EMPIN	MASMED	JUDCO
1. Assets at beginning of session	——	——					
2. Assets withdrawn on Form I	——	——					
3. Payments due for product (Totals from Form V)	——	——					
4. Net assets in bank	——	——					
5. Support cards turned in			——	——	——	——	
6. Starting income [a]			——	——			
7. Support product (multiply line 5 by line 6)			——	——			
8. Total population (TP) = _____ (divide line 7 by TP)			——	——			
9. Support multiplier			2.5	2.5	2	2	
10. Basic income [b]	——	——	——	——	——	——	——
11. National Indicator multiplier [c] (____%)							
12. Net Income (line 10 × line 11)	——	——	——	——	——	——	——

[a] For POP & SOP: Size Level One = $40
Two = $60
Three = $80

[b] Basic Income:
For BASIN/RETSIN = 10% of line 4
For POP/SOP = line 8 × line 9
For EMPIN/MASMED = line 5 × line 9
For JUDCO = Size Level One = $30
Two = $45
Three = $60

[c] Use new National Indicator levels from Form W and Table 2 from Form P to obtain correct figure.

Basic Group Income Calculations

Form X

Session # _____

	BASIN	RETSIN	POP	SOP	EMPIN	MASMED	JUDCO
1. Assets at beginning of session	——	——					
2. Assets withdrawn on Form I	——	——					
3. Payments due for product (Totals from Form V)	——	——					
4. Net assets in bank	——	——					
5. Support cards turned in			——	——	——	——	
6. Starting income [a]			——	——			
7. Support product (multiply line 5 by line 6)			——	——			
8. Total population (TP) = _____ (divide line 7 by TP)			——	——			
9. Support multiplier			2.5	2.5	2	2	
10. Basic income [b]	——	——	——	——	——	——	——
11. National Indicator multiplier [c] (____%)							
12. Net Income (line 10 × line 11)	——	——	——	——	——	——	——

[a] For POP & SOP: Size Level One = $40
                           Two = $60
                         Three = $80

[b] Basic Income:
For BASIN/RETSIN = 10% of line 4
For POP/SOP = line 8 × line 9
For EMPIN/MASMED = line 5 × line 9
For JUDCO = Size Level One = $30
                      Two = $45
                      Three = $60

[c] Use new National Indicator levels from Form W and Table 2 from Form P to obtain correct figure.

Basic Group Income Calculations

Session # _____

	BASIN	RETSIN	POP	SOP	EMPIN	MASMED	JUDCO
1. Assets at beginning of session	———	———					
2. Assets withdrawn on Form I	———	———					
3. Payments due for product (Totals from Form V)	———	———					
4. Net assets in bank	———	———					
5. Support cards turned in			———	———	———	———	
6. Starting income [a]			———	———			
7. Support product (multiply line 5 by line 6)			———	———			
8. Total population (TP) = _____ (divide line 7 by TP)			———	———			
9. Support multiplier			2.5	2.5	2	2	
10. Basic income [b]	———	———	———	———	———	———	———
11. National Indicator multiplier [c] (_____%)							
12. Net Income (line 10 × line 11)	———	———	———	———	———	———	———

[a] For POP & SOP: Size Level One = $40
                     Two = $60
                    Three = $80

[b] Basic Income:
  For BASIN/RETSIN   = 10% of line 4
  For POP/SOP   = line 8 × line 9
  For EMPIN/MASMED = line 5 × line 9
  For JUDCO   = Size Level One = $30
                              Two = $45
                              Three = $60

c Use new National Indicator levels from Form W and Table 2 from Form P to obtain correct figure.

Special Events

a. Massive Foreign Aid

    (   ) Nation A's offer accepted   (   ) Nation B's offer accepted   (   ) Nation C's offer accepted

    (   ) No action taken

b. Expeditionary Force

    (   ) No action taken   (   ) Expenditionary force raised

    (If expenditionary force is raised, p of success = 0.5)

        (   ) Successful   (   ) Unsuccessful

c. Epidemic

    <u>Exposed (names)</u>   <u>Immunized</u>   <u>Deaths (p = 0.2)</u>

d. Earthquake

    (   ) Damage repaired   (   ) Not repaired

e. Foreign Threat

    (   ) No action taken   (   ) Defence force raised

    (If defense force is raised, p of success = 0.8)

        (   ) Successful   (   ) Unsuccessful

Special Events

a. Massive Foreign Aid

   (   ) Nation A's offer accepted   (   ) Nation B's offer accepted   (   ) Nation C's offer accepted

   (   ) No action taken

b. Expeditionary Force

   (   ) No action taken   (   ) Expenditionary force raised

   (If expeditionary force is raised, p of success = 0.5)

   (   ) Successful   (   ) Unsuccessful

c. Epidemic

   <u>Exposed (names)</u>   <u>Immunized</u>   <u>Deaths ($p = 0.2$)</u>

d. Earthquake

   (   ) Damage repaired   (   ) Not repaired

e. Foreign Threat

   (   ) No action taken   (   ) Defence force raised

   (If defense force is raised, p of success = 0.8)

   (   ) Successful   (   ) Unsuccessful

Special Events

a. Massive Foreign Aid

    (   ) Nation A's offer accepted   (   ) Nation B's offer accepted   (   ) Nation C's offer accepted

    (   ) No action taken

b. Expeditionary Force

    (   ) No action taken   (   ) Expeditionary force raised

    (If expeditionary force is raised, p of success = 0.5)

        (   ) Successful   (   ) Unsuccessful

c. Epidemic

Exposed (names)	Immunized	Deaths (p = 0.2)

d. Earthquake

    (   ) Damage repaired   (   ) Not repaired

e. Foreign Threat

    (   ) No action taken   (   ) Defence force raised

    (If defense force is raised, p of success = 0.8)

        (   ) Successful   (   ) Unsuccessful

Massive Foreign Aid Offer

Three powerful foreign nations (A, B, and C), noticing the current precarious state of SIMSOC, have made offers of massive foreign aid but with strings attached. The offer from each nation is the same: a single investment of $125 for Research and Conservation and $250 for Welfare Services, if and only if their conditions are agreed to.

Nation A's Conditions

1. BASIN and RETSIN will become the property of Nation A, including all their assets.

2. There will be no head of BASIN or RETSIN within SIMSOC, but each will have a hired manager who will receive a salary of $10 per session. This manager cannot withdraw assets. The manager will be removed from office if his company fails to increase its assets in any session, and a new manager will be appointed. These managers will be chosen by the coordinator according to any criteria he wishes to use.

3. BASIN and RETSIN will each receive an allotment of five full passages or anagrams per session. All assets produced by solving these anagrams remain the property of Nation A.

Nation B's Conditions

1. All present heads of groups who live outside the Red Region will be replaced. The coordinator will appoint people who live in the Red Region as new heads of these groups.

2. The newly appointed group heads cannot be removed by the employees they hire, even by unanimous consent. If a head resigns to appoint a replacement, the replacement must live in the Red Region. A group head may be removed for failure to provide subsistence, arrest, or absence, but, in such a case, the new head appointed as a replacement will always be a member of the Red Region. A given member of the Red Region may be the head of more than one group.

Nation C's Conditions (applicable only in societies using the Minority Group Option)

1. All minority-group members who are heads of groups or agencies will be immediately removed from their positions and replaced by the standard procedures. Minority-group members may continue to work as employees.

2. As an example, two minority-group members will be placed under arrest. Any two may be chosen, but all confiscated possessions will be turned over to the bank for Nation C. If a majority of non-minority-group members is unable to come to an agreement by the end of the session, about which minority-group members should be arrested, the coordinator will choose two minority-group members at random for this purpose.

Accepting an Offer

To accept an offer, a majority of the members present must sign a petition reading, "We accept the offer of Nation _____ ." Participants may accept none of the offers, all of them, or any combination of them. The offers remain in effect through the remainder of this session and the following session. If any National Indicator is below zero, at least one offer must be accepted by the end of the session.

Expeditionary Force

A situation has developed abroad which offers opportunities to SIMSOC. Any individual or group of individuals can raise money to create an expeditionary force at any time during this or the next session. The cost of this expeditionary force is (depending on the income level of the society):

Level One: $60
Level Two: $80
Level Three: $100

This expeditionary force is created by any individual or group presenting the coordinator with the appropriate amount of money for this purpose. Such a force, if successful, will intervene in a foreign situation in ways that increase the National Indicators. If an expeditionary force is created but is unsuccessful, the National Indicators will decrease. If this option is not exercised, the National Indicators will be unaffected.

Probability of Success. An expeditionary force has a 50-50 chance of success.

Success has the following effect on the National Indicators:

	FES	SL	SC	PC
Level One:	+40	+30	+30	+30
Level Two:	+50	+40	+40	+40
Level Three:	+60	+50	+50	+50

Failure has the following effect on the National Indicators:

	FES	SL	SC	PC
Level One:	−40	−30	−30	−30
Level Two:	−50	−40	−40	−40
Level Three:	−60	−50	−50	−50

Epidemic in the Red Region

A highly contagious disease, called Red Fever, has broken out in the Red Region. Anyone exposed to this disease who is not immunized by the end of the session, has a 1 in 5 chance of dying.

Exposure. Everyone in the Red Region is exposed. In addition, anyone visiting the Red Region is exposed. Any exposed person who has not been immunized exposes everyone in any other region he enters. Thus, if someone visits Red from Yellow, then returns to Yellow without being immunized, everyone in Yellow is now exposed.

Immunization. Immunization can be obtained from the coordinator at any time during the session at a cost of $10 for each person immunized.

Deaths have the usual effect on the National Indicators. The epidemic lasts only for this session.

Earthquake

A major earthquake has disrupted transportation to SIMSOC, closing travel between some regions. You may think of the regions of SIMSOC as connected by six routes:

1. Between Green and Red
2. Between Green and Blue
3. Between Green and Yellow
4. Between Red and Blue
5. Between Red and Yellow
6. Between Blue and Yellow

The earthquake has closed routes 2, 3, 4, and 5, but routes 1 and 6 remain open. Unless routes are repaired, travel can take place only between Green and Red and between Blue and Yellow but not across these pairs of regions.

The cost of the repair is (depending on the income level of the society):

Level One:	$50
Level Two:	$75
Level Three:	$100

The roads are repaired if any individual or group presents the coordinator with the appropriate amount of money for this purpose.

If unrepaired, the damage remains for all future sessions of SIMSOC. An individual who was visiting another region at the time of the earthquake cannot return to his home region—even at the beginning of the next session—unless the route is an open one.

The National Broadcasting System is not affected by the earthquake.

Foreign Threat

Another society threatens to seize certain valuable territory belonging to SIMSOC. The members of SIMSOC have two options:

1. They may take no action to deal with this threat. Failure to act will have the following effects on the National Indicators (depending on the income level of the society):

	FES	SL	SC	PC
Level One:	−18	−9	−9	−9
Level Two:	−24	−12	−12	−12
Level Three:	−30	−15	−15	−15

2. Any individual or group may raise a defense force to resist the threat by presenting the coordinator with the appropriate amount of money for this purpose. It must be done during this session. The cost of the force is:

Level One:	$90
Level Two:	$120
Level Three:	$150

The defense force has 4 chances in 5 of being successful and rebuffing the threat. Success has the following effect on the National Indicators:

	FES	SL	SC	PC
Level One:	0	0	+3	+3
Level Two:	0	0	+4	+4
Level Three:	0	0	+5	+5

If the defense force is unsuccessful, it has the following effect on the National Indicators:

	FES	SL	SC	PC
Level One:	−18	−9	−21	−21
Level Two:	−24	−12	−28	−28
Level Three:	−30	−15	−35	−35

Coordinator's Questionnaire   Form AA

This form will enable me to make improvements in future revisions of SIMSOC. I will be grateful if you will fill it out and mail it to:

William A. Gamson
Sociology Department
University of Michigan
Ann Arbor, Michigan 48104

1. What purpose did you have in running SIMSOC?

2. Would you please describe the situation(s) in which you ran SIMSOC:
   2.1 Setting—that is, school, name of course, level, and so forth (e.g., upper-division undergraduate course in Deviance and Social Control at the University of Michigan; or, city officials and other assorted adults in workshop in Baltimore, Maryland; or, at workshop on game simulation for high school teachers).

   2.2 Size of SIMSOCs (e.g., ran two SIMSOCs, one with 38 and one with 35 participants).

   2.3 Space (e.g., four small rooms in close proximity; two adjacent rooms).

   2.4 Time schedule (e.g., in regular class sessions, MWF at 11:00 over two-week period; or, Tuesday and Thursday evening from 7:00 to 10:00 over two-week period; or, over a Saturday and Sunday, 9:00-5:00 each day).

3. Did you find any ambiguities in the rules of the game? If so, what were they?

4. Were there problems that arose in the course of play that diminished the value of the game as a learning experience? If so, what were they?

5. If you used the readings provided in the *Participant's Manual,* which ones did you find most useful? Which would you advise omitting in future editions?

6. Are there articles or exerpts from books that you feel it would be useful to add to the *Participant's Manual?*

7. How helpful did you find the *Coordinator's Manual?* How might it be made more helpful?

8. Did you find the forms and materials supplied in the *Coordinator's Manual* adequate and convenient to use? Do you have suggestions for improving them?

9. Do you have any suggestions for additional study questions or written work based on SIMSOC?

10. Did you carry out any attempts to evaluate the impact of SIMSOC on the participants? If so, please describe them.

11. What is your overall evaluation of SIMSOC? Do you intend to use it in the future? Do you have any additional suggestions not covered by the above questions?

Name: _____   Date: _____

Address: _____   May I quote you by name in future
                                                                      editions of this manual?
_____   Yes   No

Regional Summary Sheet

Form R

To: Green Region

Region	Group Heads Living There	Agency Heads Living There
Green	BASIN, JUDCO, POP	____ subsistence, ____ travel
Yellow	RETSIN, SOP	____ subsistence, ____ travel
Blue	MASMED, EMPIN	____ subsistence, ____ travel
Red	None	None

Regional Summary Sheet

Form R

To: Yellow Region

Region	Group Heads Living There	Agency Heads Living There
Green	BASIN, JUDCO, POP	____ subsistence, ____ travel
Yellow	RETSIN, SOP	____ subsistence, ____ travel
Blue	MASMED, EMPIN	____ subsistence, ____ travel
Red	None	None

Regional Summary Sheet

Form R

To: Blue Region

Region	Group Heads Living There	Agency Heads Living There
Green	BASIN, JUDCO, POP	____ subsistence, ____ travel
Yellow	RETSIN, SOP	____ subsistence, ____ travel
Blue	MASMED, EMPIN	____ subsistence, ____ travel
Red	None	None

Regional Summary Sheet

Form R

To: Red Region

Region	Group Heads Living There	Agency Heads Living There
Green	BASIN, JUDCO, POP	____ subsistence, ____ travel
Yellow	RETSIN, SOP	____ subsistence, ____ travel
Blue	MASMED, EMPIN	____ subsistence, ____ travel
Red	None	None

Regional Summary Sheet

Form R

To: Green Region

Region	Group Heads Living There	Agency Heads Living There
Green	BASIN, JUDCO, POP	___ subsistence, ___ travel
Yellow	RETSIN, SOP	___ subsistence, ___ travel
Blue	MASMED, EMPIN	___ subsistence, ___ travel
Red	None	None

Regional Summary Sheet

Form R

To: Yellow Region

Region	Group Heads Living There	Agency Heads Living There
Green	BASIN, JUDCO, POP	___ subsistence, ___ travel
Yellow	RETSIN, SOP	___ subsistence, ___ travel
Blue	MASMED, EMPIN	___ subsistence, ___ travel
Red	None	None

Regional Summary Sheet

Form R

To: Blue Region

Region	Group Heads Living There	Agency Heads Living There
Green	BASIN, JUDCO, POP	___ subsistence, ___ travel
Yellow	RETSIN, SOP	___ subsistence, ___ travel
Blue	MASMED, EMPIN	___ subsistence, ___ travel
Red	None	None

Regional Summary Sheet

Form R

To: Red Region

Region	Group Heads Living There	Agency Heads Living There
Green	BASIN, JUDCO, POP	___ subsistence, ___ travel
Yellow	RETSIN, SOP	___ subsistence, ___ travel
Blue	MASMED, EMPIN	___ subsistence, ___ travel
Red	None	None

(1)   In May, 1937, shortly before Memorial Day, 78,000 steelworkers began a strike against the "Little Steel" companies of Bethlehem Steel, Republic Steel, Inland Steel, and Youngstown Sheet and Tube. The CIO-backed Steel Workers Organizing Committee (SWOC) was less than a year old at the time but had already enjoyed some notable success.

---

(2)   A few months earlier, it had signed collective-bargaining agreements with the five largest U.S. Steel subsidiaries, and, by early May, SWOC had signed contracts with 110 firms.

---

(3)   The Little Steel companies, however, were prepared to resist. Under the leadership of Tom M. Girdler, president of Republic Steel, they refused to sign an agreement which they felt, in Girdler's words, "was a bad thing for our companies, for our employees; indeed, for the United States of America" (Galenson, 1960, p. 96).

---

(4)   The decision to resist was made more ominous by the common practice of large employers of the time to stock arsenals of weapons and tear gas in anticipation of labor disputes. Much of our information comes from the report of the LaFollette Committee of the United States Senate, which investigated the events surrounding the Little Steel strike.

(5)   The committee report noted, for example, that during the years 1933 to 1937, over a million dollars' worth of tear gas and sickening gas was purchased by employers and law-enforcement agencies, but that "all of the largest individual purchasers are corporations and that their totals far surpass those of large law-enforcement purchasers" (quoted in Sweeney, 1956, p. 20).

(6)   The largest purchaser of gas equipment in the country was none other than the Republic Steel Corporation, which "bought four times as much as the largest law-enforcement purchaser." The Republic Steel aresenal included 552 revolvers, 61 rifles with 1325 rounds of ammunition, and 245 shotguns in addition to gas grenades (Sweeney, p. 33)

(7)   The Little Steel strike began on May 26, 1937, and for a few days prior to May 30, picketing and arrests occurred near Republic Steel's mill in south Chicago. On Memorial Day, after a mass meeting at strike headquarters, the strikers decided to march to the plant to establish a mass picket line. A crowd of about 1,000 persons, "headed by two bearers of American Flags,...started across the prairie toward the street which fronts on the mill. There was a holiday spirit over the crowd" (Sweeney, p. 33).

(8)   Take the Brotherhood of the Cooperative Commonwealth, for example. Born in the ferment of the 1890s, it was the brainchild of an obscure Maine reformer, Norman Wallace Lermond. "Its immediate and most important objective was to colonize *en masse* a sparsely inhabited Western state with persons desiring to live in socialist communities. Once established, the colonists would be in a position to capture control of the state's government and lay the foundation for a socialist commonwealth" (Quint, 1964).

(9)   Not much happened for the first year of its existence, but Lermond was "a letter-writing dynamo and he bombarded reformers throughout the country with appeals for assistance." He began to get some results. Imogene C. Fales, a New York reformer, "who was a charter member of innumerable humanitarian and socialist movements in the 1880s and 1890s, agreed to serve with Lermond as co-organizer" (Quint).

----

(10)   But the big catch for the fledgling challenging group was Eugene V. Debs. Debs had been recently released from his prison term for defying the injunction against the American Railway Union which broke the Pullman strike. He was a genuine hero of the left, who was now, for the first time, espousing socialism. Debs was a thoroughly decent person who lacked the vituperative personal style so characteristic of many leftists.

----

(11)   Furthermore, he was an extraordinarily effective platform speaker where, as Quint describes him, "the shining sincerity of his speeches and the flowing honesty of his personality more than compensated for the lack of knowledge of the more delicate points of Marxist theory. His soul was filled with a longing for social justice and he communicated this feeling to the audiences who gathered to hear him extol the new Social Democracy."

----

(12)   Debs became attracted to the colonization scheme. "Give me 10,000 men," Debs told a socialist convention, "aye, 10,000 in a western state with access to the sources of production, and we will change the economic conditions, and we will convince the people of that state, win their hearts and their intelligence. We will lay hold upon the reins of government and plant the flag of Socialism upon the State House" (quoted in Quint).

(13)   There is a certain absurdity in comparing a group that seeks a modest change and threatens no major redistribution of power with one that seeks to sweep aside the old order and all its supporting institutions. For modesty of aspiration, few challenging groups can compete with the Society for the Promotion of Manual Labor in Literary Institutions.

(14)   This quaint effort of the 1830s was one of several reform efforts supported by the Tappan brothers. The group's mobilization effort consisted primarily of lectures by its paid general agent, Theodore Weld. No state militias were necessary to hold back the crowds aroused by Weld's impassioned pleas. In fact, Weld, who was apparently a skillful speaker, frequently found it expedient to build his audience by advertising his topic as temperance, then using the occasion to make an additional pitch on educational reform.

(15)   Weld invoked an image of the college student and seminarian that 150 years has not seriously dated. His portrait depicted the typical collegian "with his feet elevated upon a mantelpiece as high as his head, body bent like a halfmoon or a horseshoe, lolling, stretching, yawning, smoking, snoring."

(16)   Strong physical labor was recommended for this shiftless lot, to "tone the body, stimulate the intellect, safeguard the student's morals by occupying his spare time, teach him useful skills, promote industry, originality, and manliness. By cheapening the cost of education, it would broaden the country's intellectual base, and by demonstrating the compatibility of physical and intellectual endeavor, it would do away with absurd social distinctions between those who work with their brains and those who produce with their hands" (Thomas, 1950).

(17)   In contrast, although in a way equally quaint, consider the Communist Labor Party. This was the name taken by the so-called Benjamin Gitlow-John Reed, left-wing faction of the Socialist Party when they broke away in the summer of 1919. It was a period of considerable hysteria, stimulated by the success of the Bolshevik revolution and the substantial labor unrest in the United States.

----

(18)   Nineteen-nineteen was the year of the Seattle general strike, the Lawrence textile strike, the Boston police strike, the national coal strike and a large steel strike involving 365,000 workers.

    The fledgling and ineffectual Communist Labor Party was a natural target of the various organs of the anti-red hysteria. First, in November, 1919, many of its members, including Gitlow, were arrested in raids growing out of the New York State Senate's Lusk Committee and its investigations of "seditious activity."

----

(19)   There was considerable competition among anticommunist entrepreneurs of the day; group members were also arrested in large numbers in raids initiated by the District Attorney of Chicago and U.S. Attorney General Mitchell Palmer. Gitlow also reports a great deal of additional harassment (Gitlow, 1940), and he personally was sentenced and jailed for more than a year.

----

(20)   John Reed once faced three indictments at the same time. Most of the leaders "lived in a half-world of indictments, trials, defense committees, convictions, sentences, and appeals" (Draper, 1957).

    What were the goals of this beleaguered group? "The Communist Labor Party of America declares itself in complete accord with the principles of Communism as laid down in the Manifesto of the Third International formed at Moscow."

(21)   The League of Deliverance won both acceptance and new advantages, but there is some reason to question whether displacement of its antagonist was a true goal. This nativist group of the 1880s tried to prevent the employment of Chinese labor by boycotting those businesses that continued to employ Chinese.

---

(22)   "Don't patronize Grass or Butterfield; they sell Chinese-made boots and shoes. Avoid them! They are traitors to their race," said one of the League's leaflets. However, destruction of such businesses was really more means than goal—that is, not a justification in its own right. By getting rid of its Chinese workers, a business could buy peace with the League and many did just that.

---

(23)   The National Student League, on the other hand, was equally Marxist in orientation but adopted a number of specific causes as its own. It had a series of campus-oriented concerns such as ending compulsory ROTC and compulsory chapel and gaining equal educational opportunities for women and blacks.

---

(24)   It cooperated with other groups in the early 1930s in sponsoring antiwar strikes and involved itself in the bitter struggle of the coal miners in Harlan County, Kentucky. No doubt, many National Student League participants saw these separate struggles as part of the process of building "consciousness" for the larger struggle, but they became goals in their own right as well.

(25)   Not all of the challenging groups in this study began life friendless. About one-fourth of them (14) enjoyed the patronage of some individual or group with significant power or wealth. Although these groups, by definition, were also attempting to mobilize a constituency, they were not as completely dependent on this constituency for resources as an unsponsored group was.

---

(26)   The American Committee for the Outlawry of War, for example, enjoyed the patronage of an energetic and well-connected lawyer, Salmon O. Levinson. In many ways, the committee was a vehicle for Levinson's crusade. He made sporadic efforts to raise money from others but without much success.

---

(27)   At one point, he "tried charging a small price for printed material that he had been sending out gratis, especially when large batches were called for. But when Amy Woods of the Women's International League for Peace and Freedom asked for fifty thousand pamphlets and insisted that there was no money to pay for them, he made an exception."

---

(28)   "In point of fact, exceptions were made in nearly every case, so that very little money was raised from that source" (Stoner, 1942, p. 108). Levinson did not have unlimited personal funds, and made plaintive pleas for help from rich angels but without success.

(29)   Engrossed as he was in the campaign, Levinson "really never made any but scattered and sporadic attempts to raise money. He much preferred to give what was needed out of his own pocket rather than be distracted from his main object.... He began to pay more than 95 percent of the expense of the campaign out of his own pocket, and he continued to do so even when it was costing upward of fifteen thousand dollars annually" (Stoner, pp. 108-110).

----

(30)   In addition to this kind of personal sponsorship, there is organizational sponsorship. The fledgling Steel Workers Organizing Committee did not have to depend on its nascent membership for the resources it needed to carry on its battles with the steel producers. The CIO was ready to lend its help.

----

(31)   "[John L.] Lewis had given us a check for $25,000," David McDonald explained, "[but] that barely got our office open. The task of putting together an effective group of organizers was the heart of our effort. It was also expensive, and Lewis knew it. Within a few weeks, he sent us a second check, this time for $500,000" (McDonald, 1969, p. 91).

----

(32)   When it came time to select quarters, McDonald chose a suite "on the thirty-sixth floor of the Grant Building—Pittsburgh's tallest at that time—[which] looked out splendidly over the industrial heart of the city. It also topped by several floors offices of some of the steel companies we hoped to organize and left no doubt of our permanence and stability.... The word got around quickly in the right places that this was no fly-by-night effort but a well-financed movement of labor union professionals who knew what they were about and meant business" (McDonald, p. 91).

(33)   It will, I think, be universally admitted that the average man (and it has not yet been discovered that the average wheelman is built on a different plan from the ordinary citizen) in nowise attempts to conceal his anxiety to know what he gets out of anything he is solicited to "go into," whether it be a real-estate deal, stock-jobbing operation, secret society, or beneficial or fraternal organization— and justly so.

(34)   The quid pro quo obtains in all our dealings from the cradle to the grave.... During a period extending over something like ten years of League work, my experience has almost invariably been...that the first question broached by a rider, when requested to join the L.A.W., is "What do I get out of it?" And, mind you, he is usually quite deaf to any sentimental arguments.

(35)   The benefits of fellowship with the thirty-odd thousand of us who go to make up the elect have no weight with him. He wants—and justly so, again—to see paid down to him in hand the material benefits.... This being the case, what is to do? Why, give it to him, of course.

(36)   Herein lies, in a nutshell, the secret of that future growth of the organization which can make the League of American Wheelmen that power in the land which it can become under properly directed effort.... Pennsylvania owes no small measure of her growth to the fine road books which are furnished to her members free...in addition to the weekly paper which each one of them receives.

(37)   The violence of the Night Riders was the most organized of any group studied. They "made their first show of armed force at Princeton [Kentucky], on the morning of Saturday, December 1, 1906 when shortly after midnight approximately 250 armed and masked men took possession of the city and dynamited and burned two large tobacco factories...."

----

(38)   Citizens in the business district opened windows and looked out on bodies of masked men hurrying along with guns on their shoulders. They saw other masked men and armed men patrolling the sidewalks and street corners and they heard commands: "Get back!" And if they did not obey, bullets splattered against the brick walls near by or crashed through the window panes above their heads....

----

(39)   Several squads of men had marched in along the Cadiz road and captured the police station, the waterworks plants, the courthouse, and the telephone and telegraph offices. They had disarmed the policemen and put them under guard, shut off the city water supply, and taken the places of the telephone and telegraph operators....

----

(40)   "Within a few minutes the city was in control of the Riders and all communication with the outside was cut off." With their mission accomplished and the tobacco factories in flames, the men "mounted their horses and rode away singing 'The fire shines bright in my old Kentucky home' " (Nall, 1942, p. 69).

Report to MASMED   End of Session # _____   Form Y

National Indicators	Absentees _____	Group Support	Goal Declarations	Riots
FES = _____	Unemployed _____	POP _____	Positive _____	No. of Rioters _____
SL = _____	Rioters _____	SOP _____	Negative _____	No. of Guard Posts ____
SC = _____	Arrests _____	EMPIN _____	Changed _____	
PC = _____	Deaths _____	MASMED _____		

Report to MASMED   End of Session # _____   Form Y

National Indicators	Absentees _____	Group Support	Goal Declarations	Riots
FES = _____	Unemployed _____	POP _____	Positive _____	No. of Rioters _____
SL = _____	Rioters _____	SOP _____	Negative _____	No. of Guard Posts ____
SC = _____	Arrests _____	EMPIN _____	Changed _____	
PC = _____	Deaths _____	MASMED _____		

Report to MASMED   End of Session # _____   Form Y

National Indicators	Absentees _____	Group Support	Goal Declarations	Riots
FES = _____	Unemployed _____	POP _____	Positive _____	No. of Rioters _____
SL = _____	Rioters _____	SOP _____	Negative _____	No. of Guard Posts ____
SC = _____	Arrests _____	EMPIN _____	Changed _____	
PC = _____	Deaths _____	MASMED _____		

Report to MASMED   End of Session # _____   Form Y

National Indicators	Absentees _____	Group Support	Goal Declarations	Riots
FES = _____	Unemployed _____	POP _____	Positive _____	No. of Rioters _____
SL = _____	Rioters _____	SOP _____	Negative _____	No. of Guard Posts ____
SC = _____	Arrests _____	EMPIN _____	Changed _____	
PC = _____	Deaths _____	MASMED _____		

Report to MASMED

End of Session # _____   Form Y

National Indicators	Absentees	Group Support	Goal Declarations	Riots
FES = _____	Unemployed _____	POP _____	Positive _____	No. of RiEvents _____
SL = _____	Rioters _____	SOP _____	Negative _____	No. of Guard Posts ____
SC = _____	Arrests _____	EMPIN _____	Changed _____	
PC = _____	Deaths _____	MASMED _____		

Report to MASMED

End of Session # _____   Form Y

National Indicators	Absentees	Group Support	Goal Declarations	Riots
FES = _____	Unemployed _____	POP _____	Positive _____	No. of Rioters _____
SL = _____	Rioters _____	SOP _____	Negative _____	No. of Guard Posts ____
SC = _____	Arrests _____	EMPIN _____	Changed _____	
PC = _____	Deaths _____	MASMED _____		

Report to MASMED

End of Session # _____   Form Y

National Indicators	Absentees	Group Support	Goal Declarations	Riots
FES = _____	Unemployed _____	POP _____	Positive _____	No. of Rioters _____
SL = _____	Rioters _____	SOP _____	Negative _____	No. of Guard Posts ____
SC = _____	Arrests _____	EMPIN _____	Changed _____	
PC = _____	Deaths _____	MASMED _____		

Report to MASMED

End of Session # _____   Form Y

National Indicators	Absentees	Group Support	Goal Declarations	Riots
FES = _____	Unemployed _____	POP _____	Positive _____	No. of Rioters _____
SL = _____	Rioters _____	SOP _____	Negative _____	No. of Guard Posts ____
SC = _____	Arrests _____	EMPIN _____	Changed _____	
PC = _____	Deaths _____	MASMED _____		

Report to MASMED

End of Session # _____   Form Y

National Indicators	Absentees _____	Group Support	Goal Declarations	Riots
FES = _____	Unemployed _____	POP _____	Positive _____	No. of Rioters _____
SL = _____	Rioters _____	SOP _____	Negative _____	No. of Guard Posts ____
SC = _____	Arrests _____	EMPIN _____	Changed _____	
PC = _____	Deaths _____	MASMED _____		

Report to MASMED

End of Session # _____   Form Y

National Indicators	Absentees	Group Support	Goal Declarations	Riots
FES = _____	Unemployed _____	POP _____	Positive _____	No. of Rioters _____
SL = _____	Rioters _____	SOP _____	Negative _____	No. of Guard Posts ____
SC = _____	Arrests _____	EMPIN _____	Changed _____	
PC = _____	Deaths _____	MASMED _____		

Report to MASMED

End of Session # _____   Form Y

National Indicators	Absentees	Group Support	Goal Declarations	Riots
FES = _____	Unemployed _____	POP _____	Positive _____	No. of Rioters _____
SL = _____	Rioters _____	SOP _____	Negative _____	No. of Guard Posts ____
SC = _____	Arrests _____	EMPIN _____	Changed _____	
PC = _____	Deaths _____	MASMED _____		

Report to MASMED

End of Session # _____   Form Y

National Indicators	Absentees	Group Support	Goal Declarations	Riots
FES = _____	Unemployed _____	POP _____	Positive _____	No. of Rioters _____
SL = _____	Rioters _____	SOP _____	Negative _____	No. of Guard Posts ____
SC = _____	Arrests _____	EMPIN _____	Changed _____	
PC = _____	Deaths _____	MASMED _____		

Report to MASMED End of Session # _____ Form Y

National Indicators	Absentees _____	Group Support	Goal Declarations	Riots
FES = _____	Unemployed _____	POP _____	Positive _____	No. of Rioters _____
SL = _____	Rioters _____	SOP _____	Negative _____	No. of Guard Posts ____
SC = _____	Arrests _____	EMPIN _____	Changed _____	
PC = _____	Deaths _____	MASMED _____		

Report to MASMED End of Session # _____ Form Y

National Indicators	Absentees _____	Group Support	Goal Declarations	Riots
FES = _____	Unemployed _____	POP _____	Positive _____	No. of Rioters _____
SL = _____	Rioters _____	SOP _____	Negative _____	No. of Guard Posts ____
SC = _____	Arrests _____	EMPIN _____	Changed _____	
PC = _____	Deaths _____	MASMED _____		

Report to MASMED End of Session # _____ Form Y

National Indicators	Absentees _____	Group Support	Goal Declarations	Riots
FES = _____	Unemployed _____	POP _____	Positive _____	No. of Rioters _____
SL = _____	Rioters _____	SOP _____	Negative _____	No. of Guard Posts ____
SC = _____	Arrests _____	EMPIN _____	Changed _____	
PC = _____	Deaths _____	MASMED _____		

Report to MASMED End of Session # _____ Form Y

National Indicators	Absentees _____	Group Support	Goal Declarations	Riots
FES = _____	Unemployed _____	POP _____	Positive _____	No. of Rioters _____
SL = _____	Rioters _____	SOP _____	Negative _____	No. of Guard Posts ____
SC = _____	Arrests _____	EMPIN _____	Changed _____	
PC = _____	Deaths _____	MASMED _____		

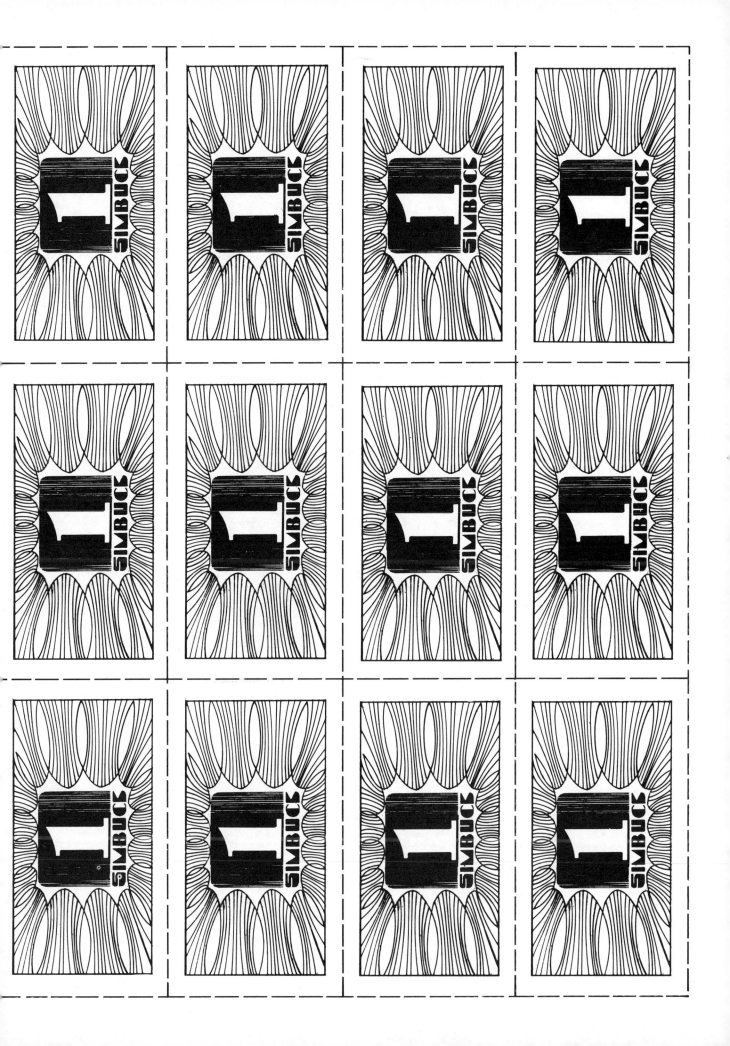

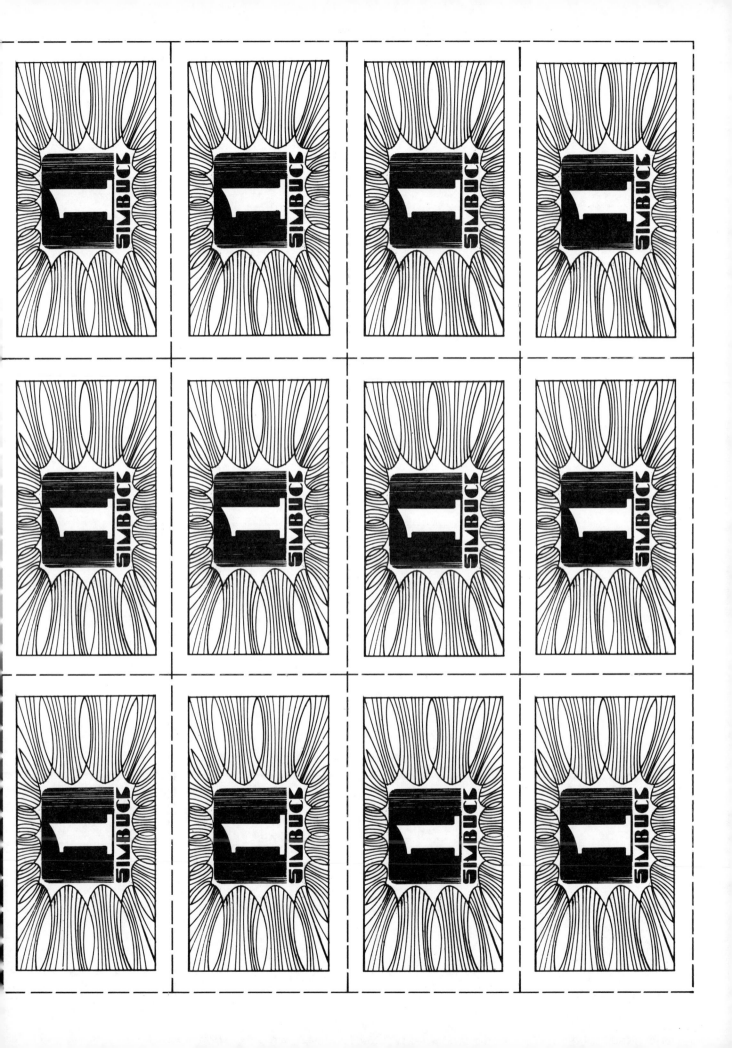

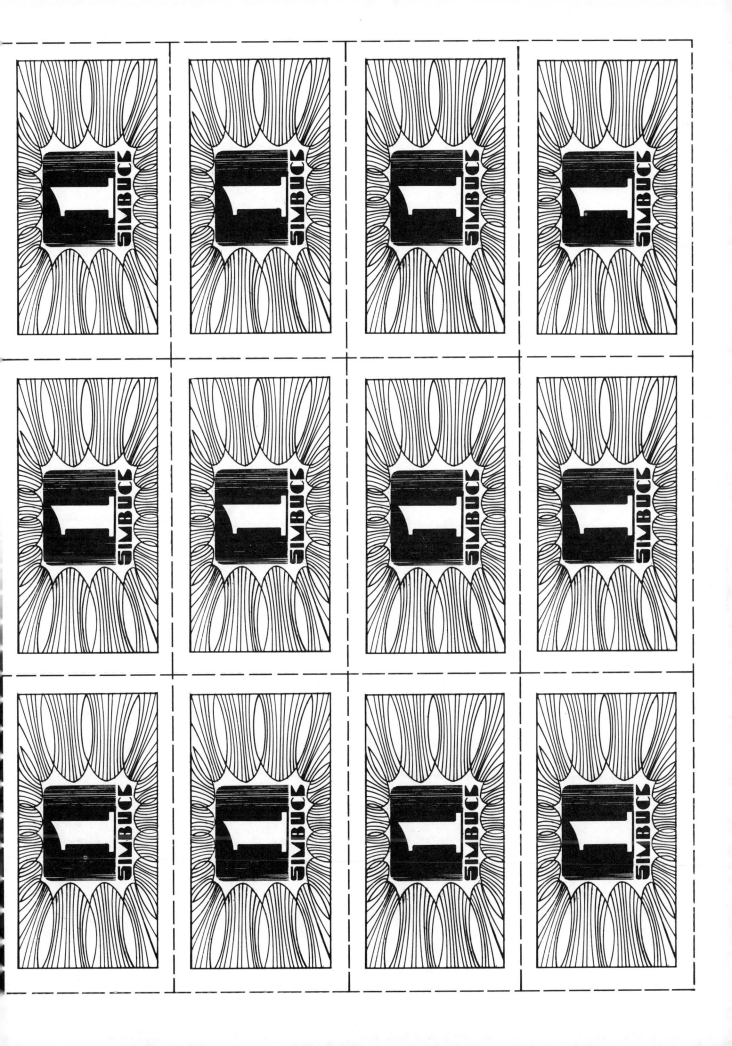

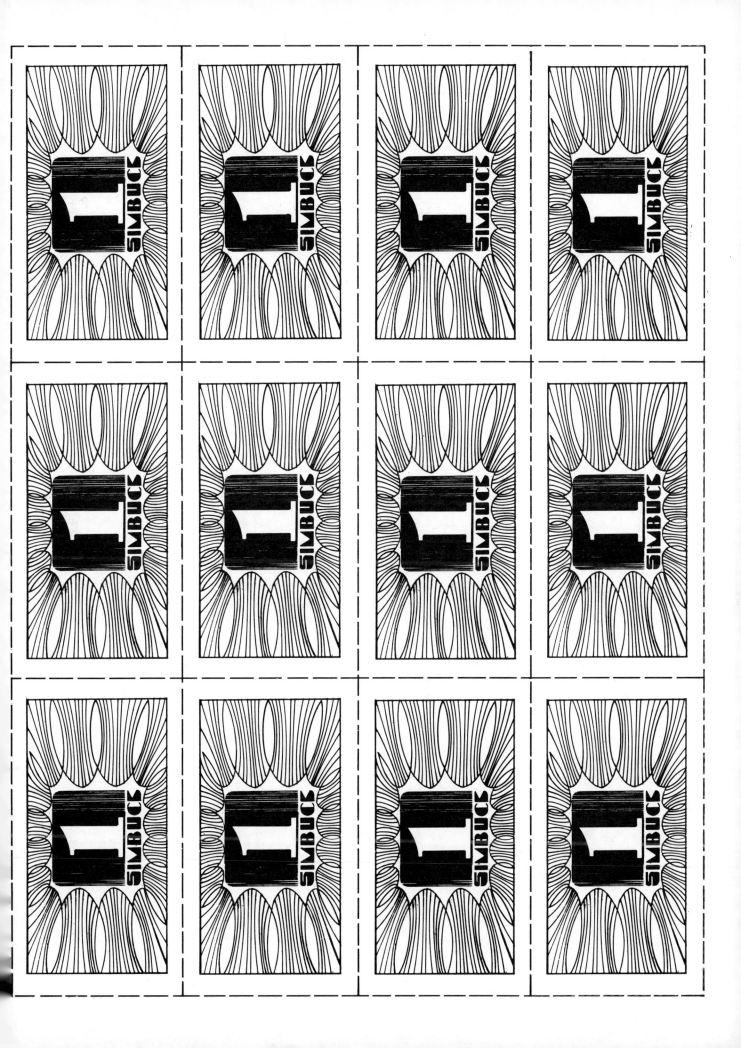

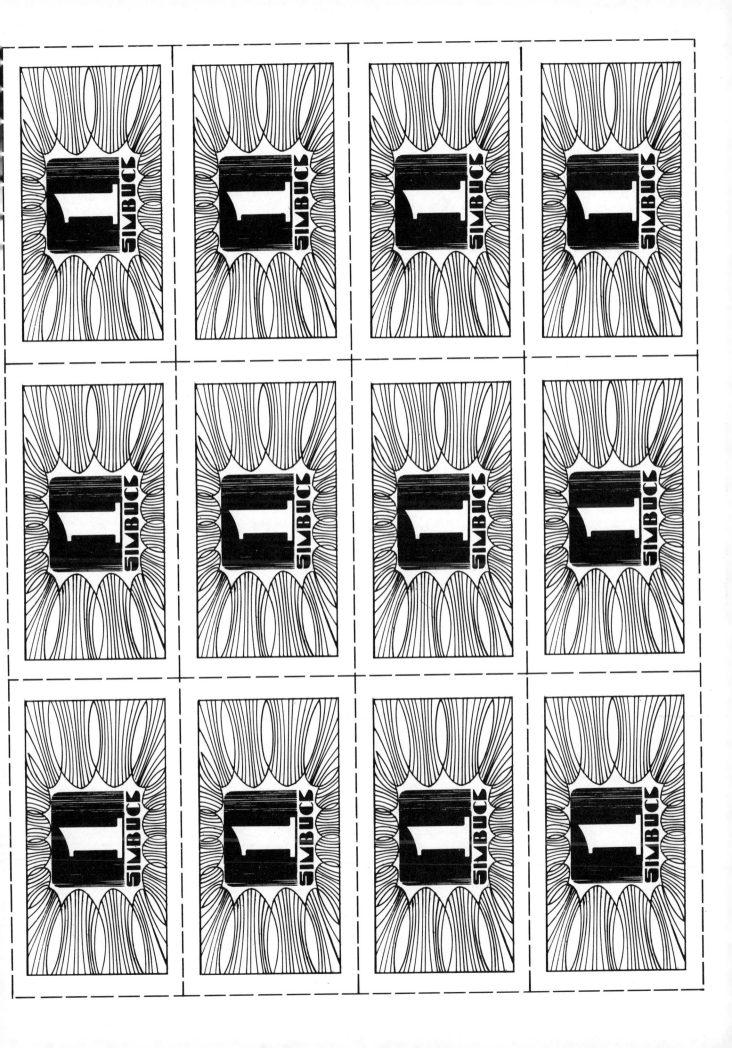

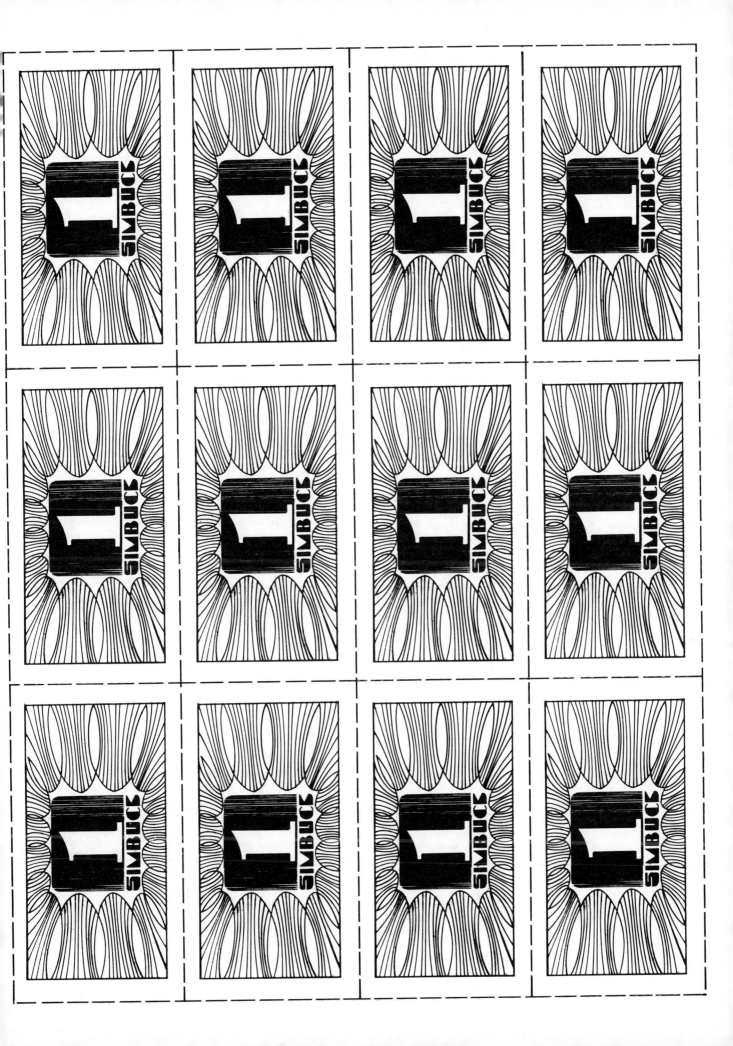

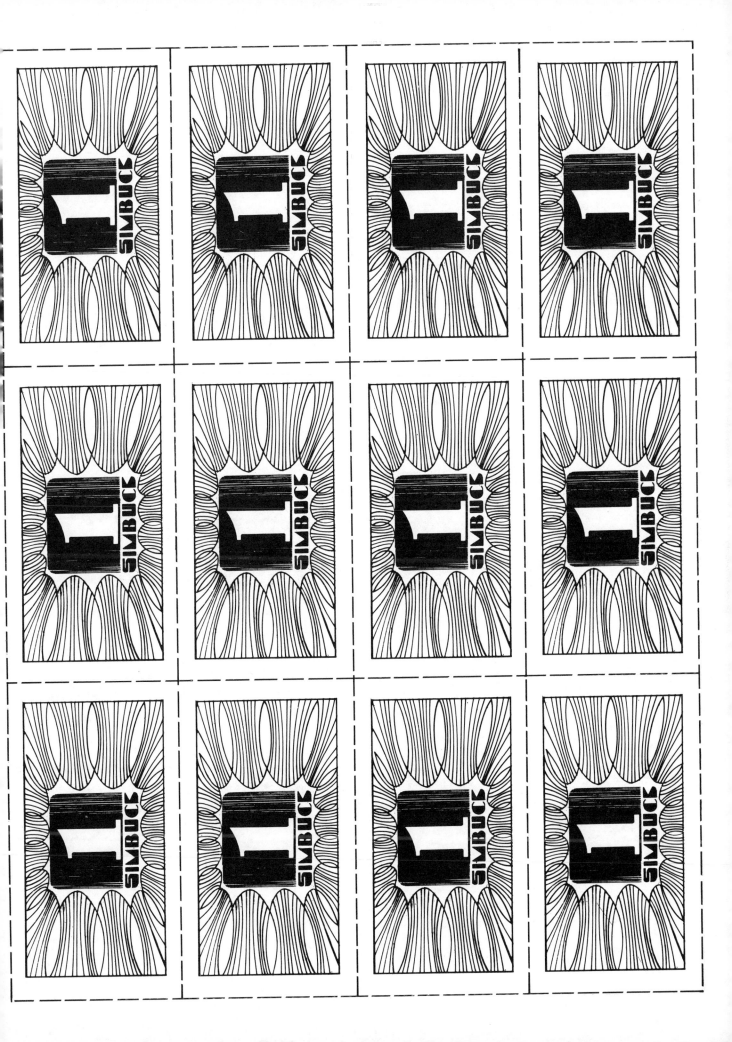

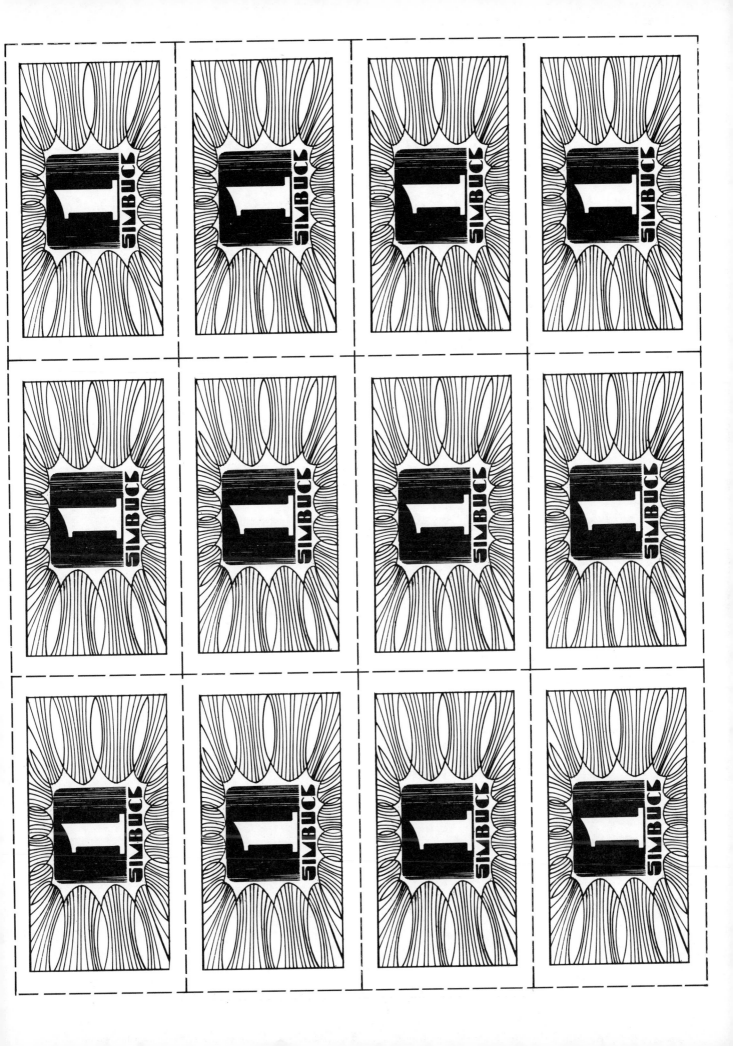

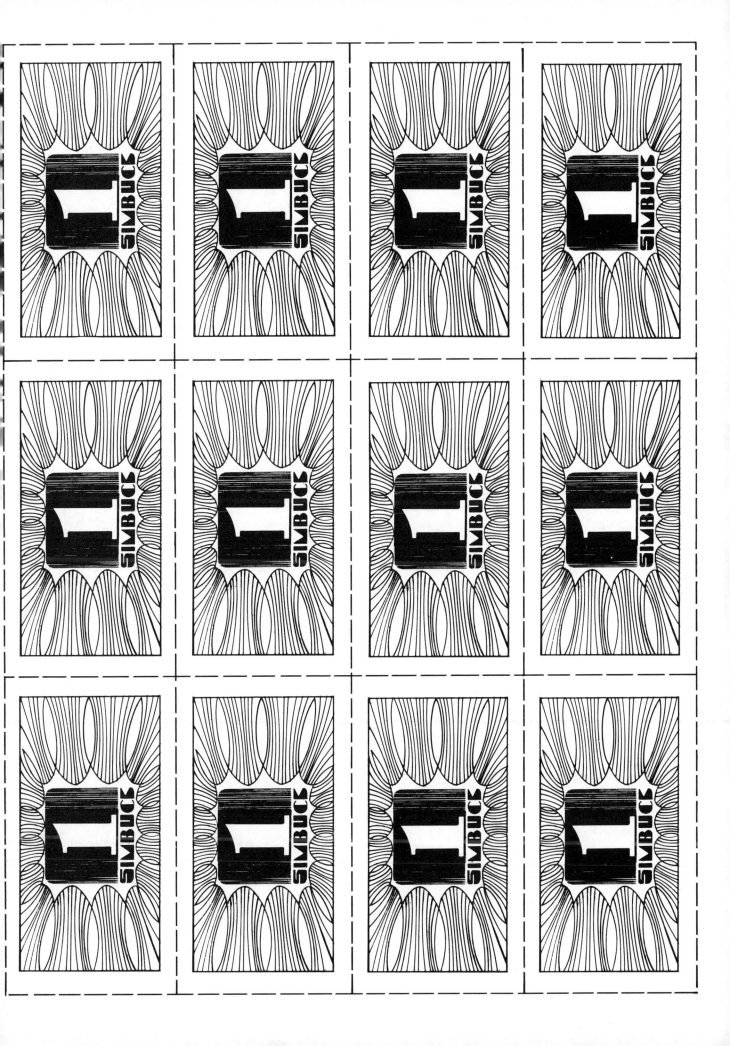

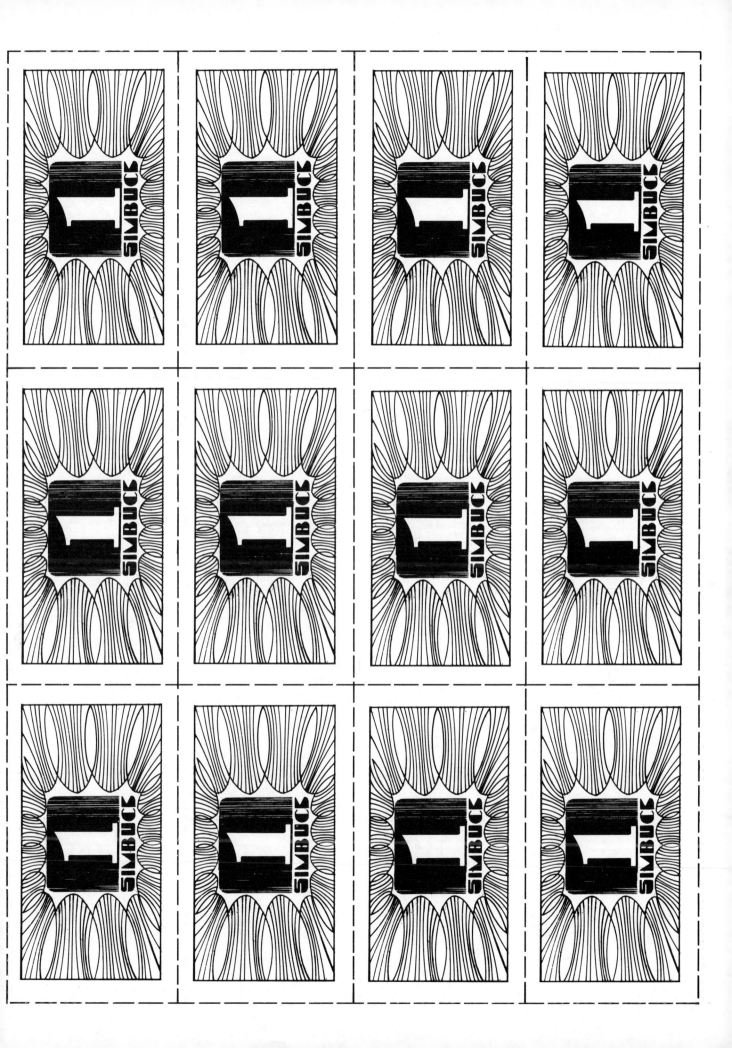

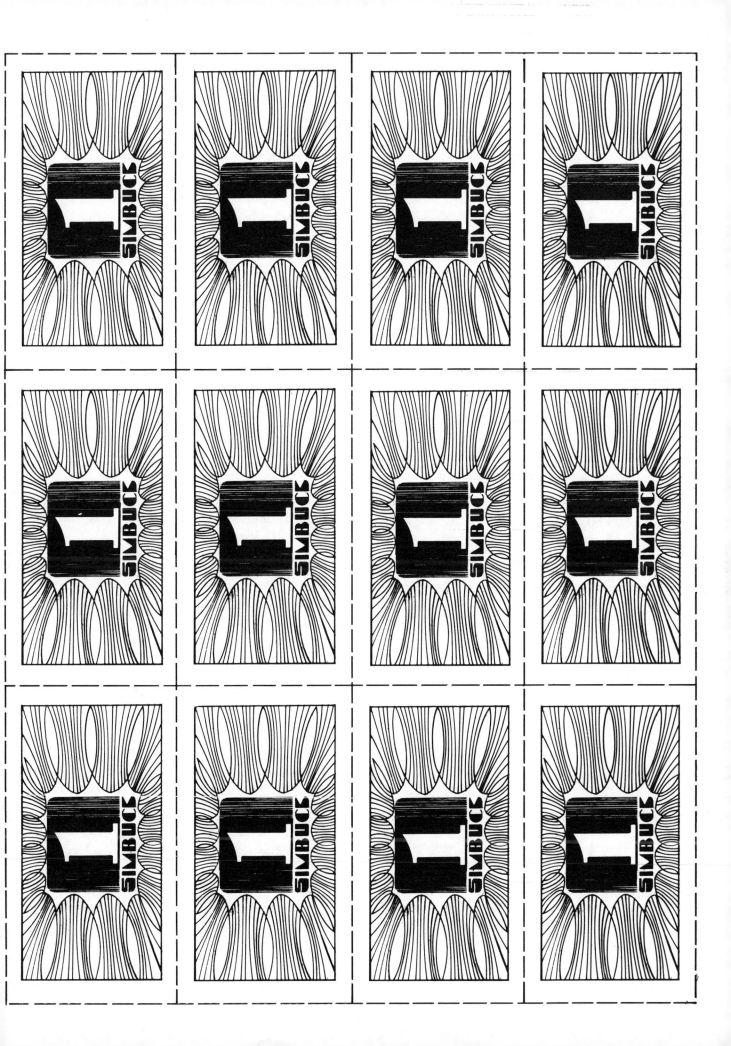

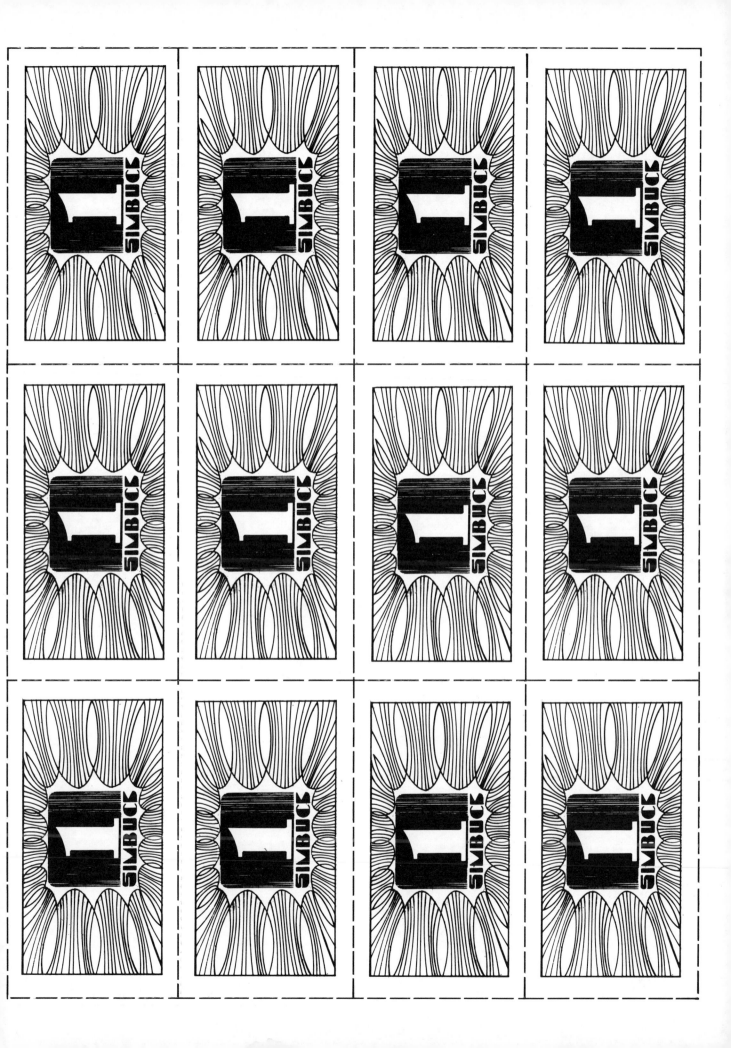

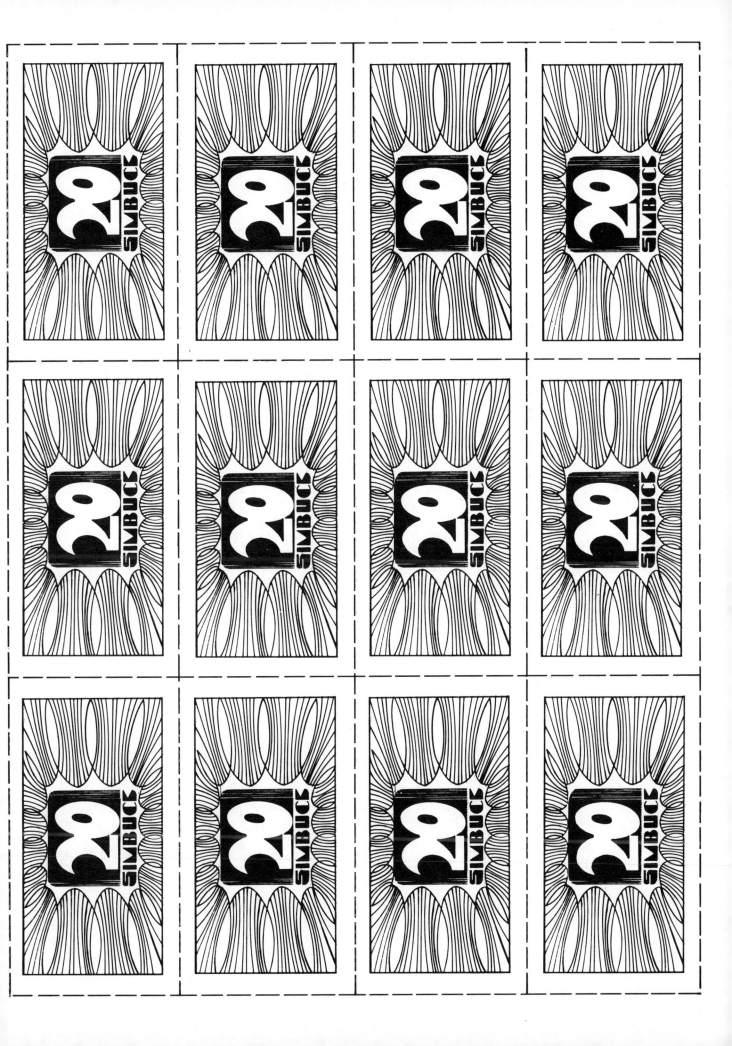

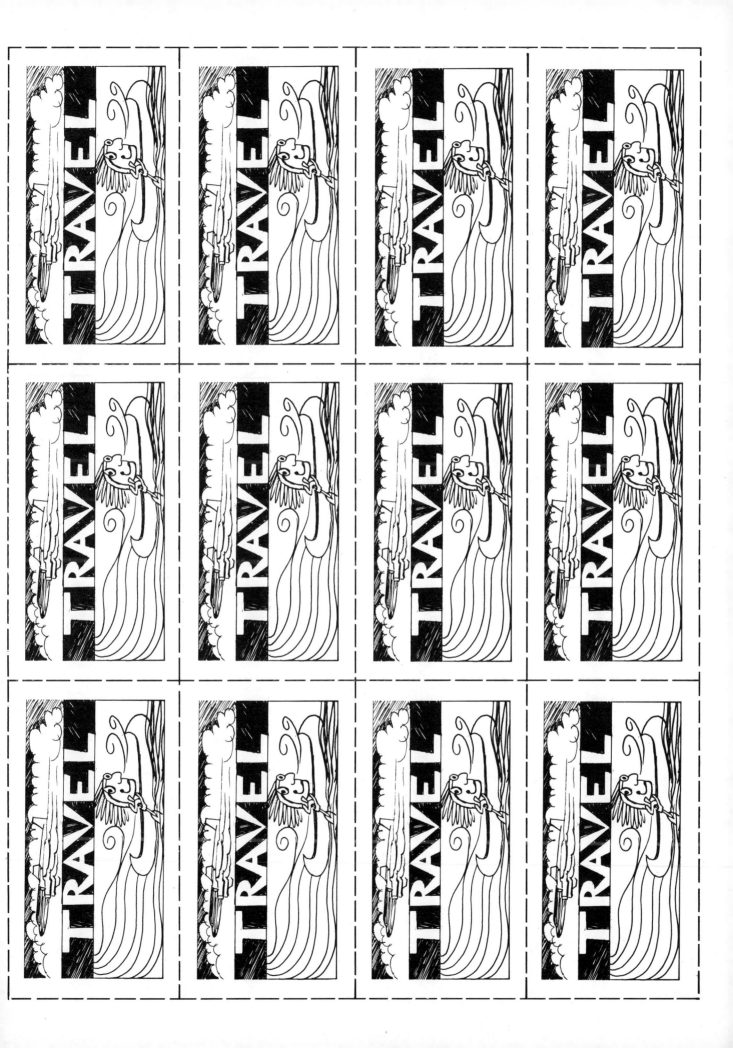

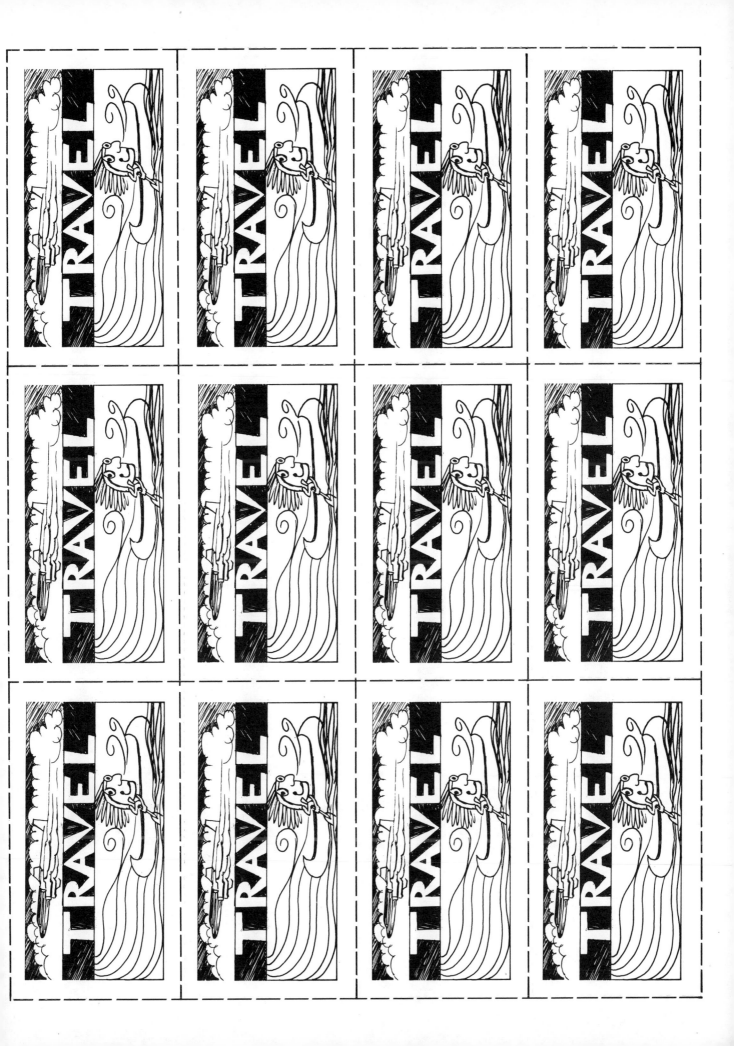

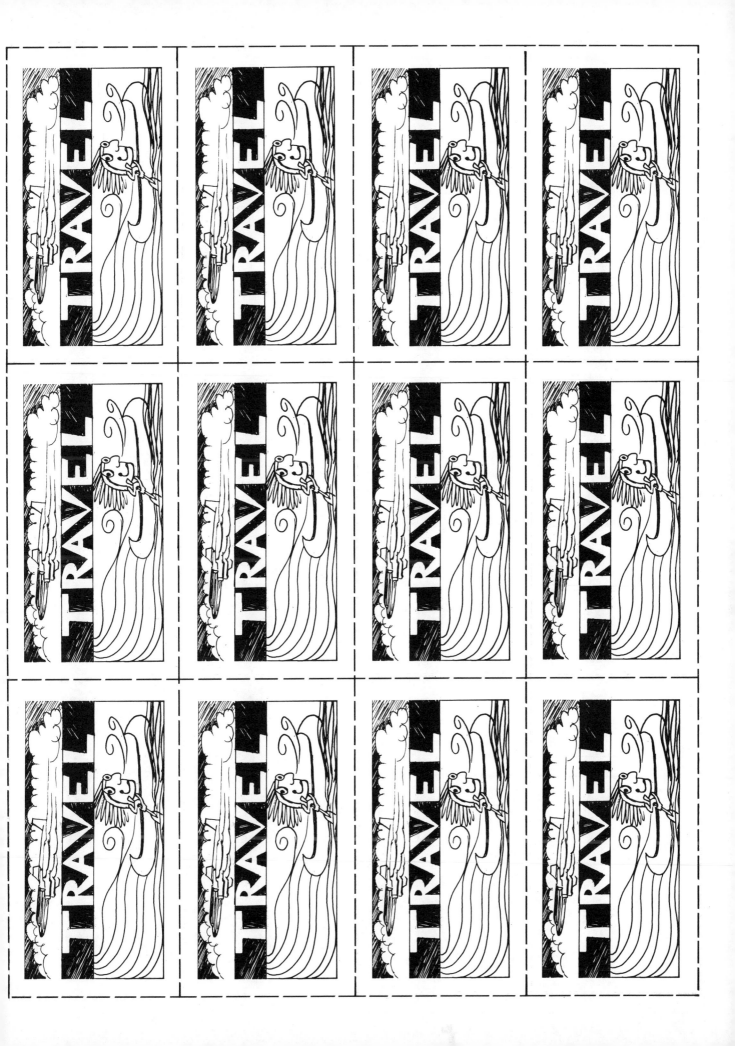

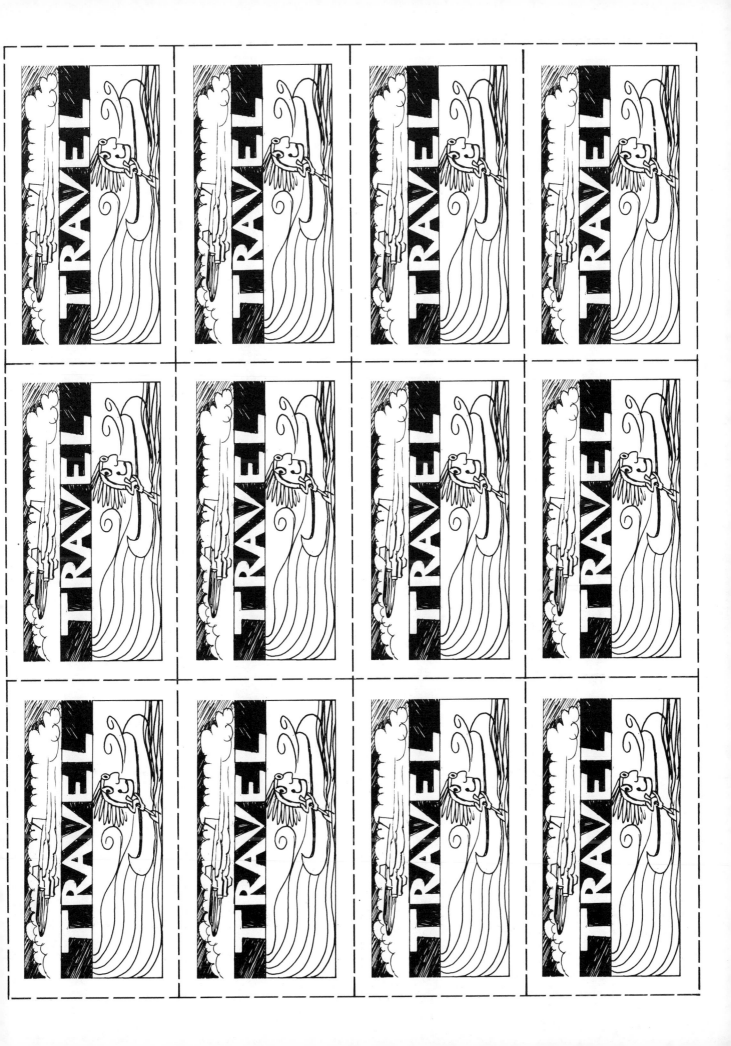

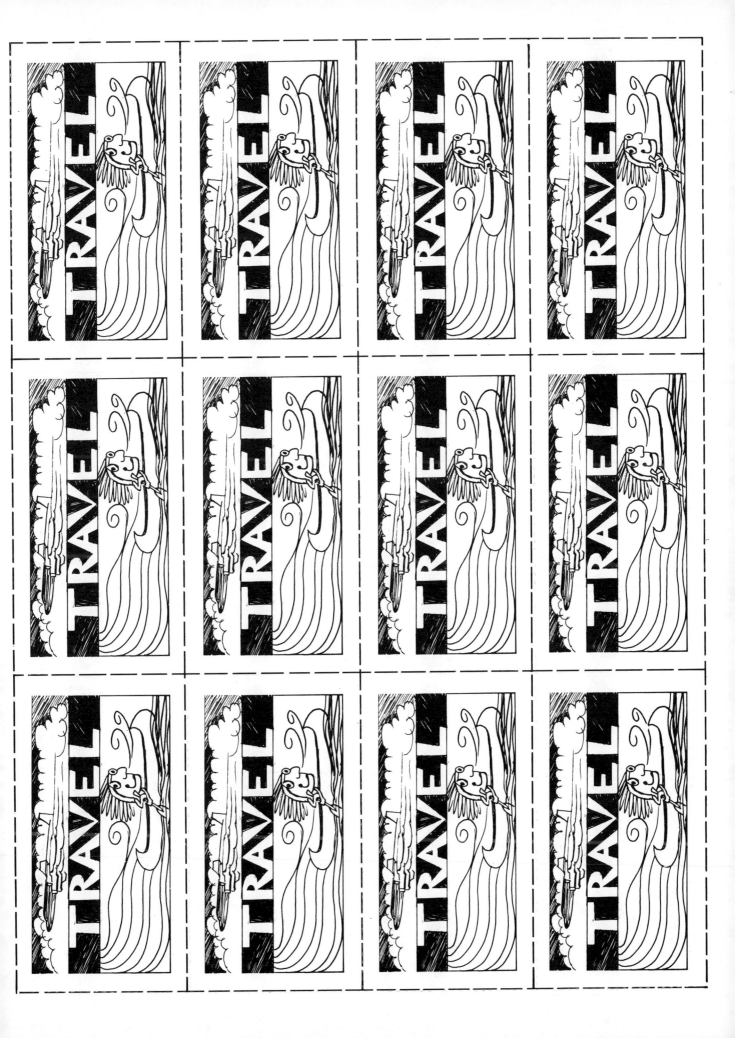

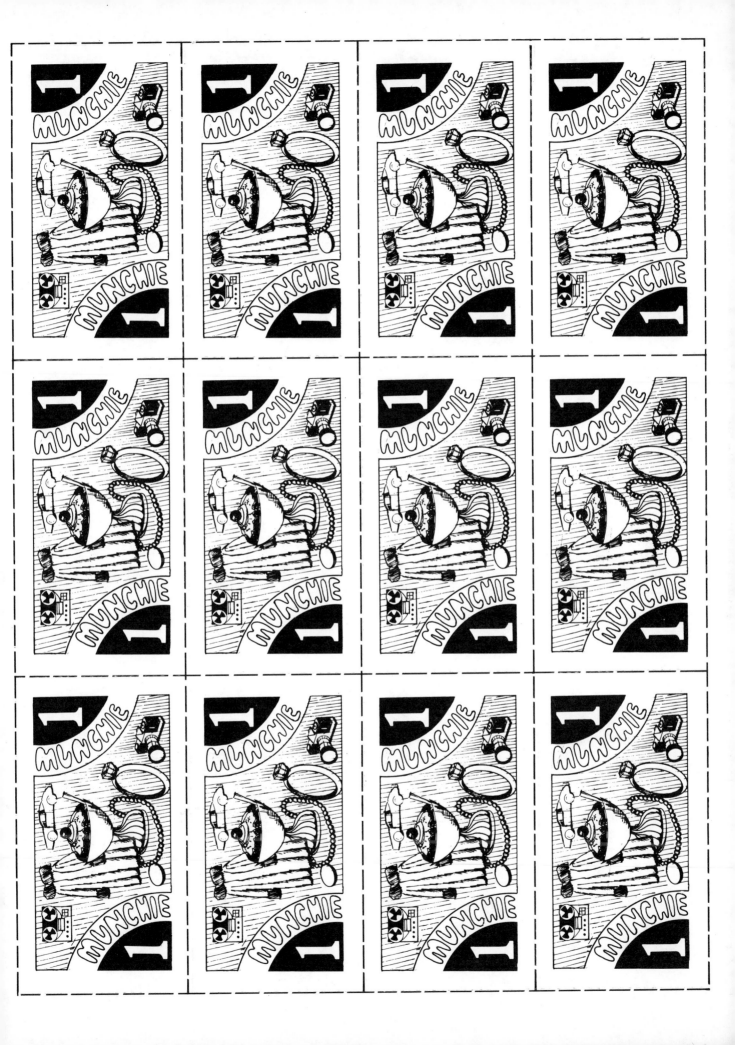

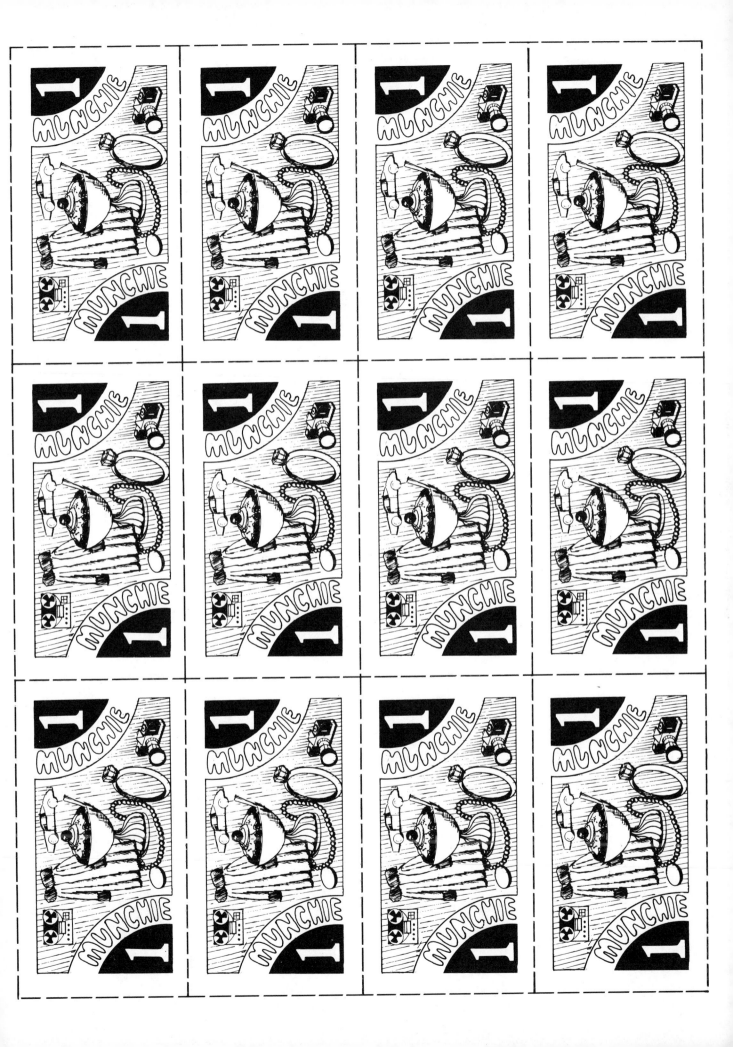

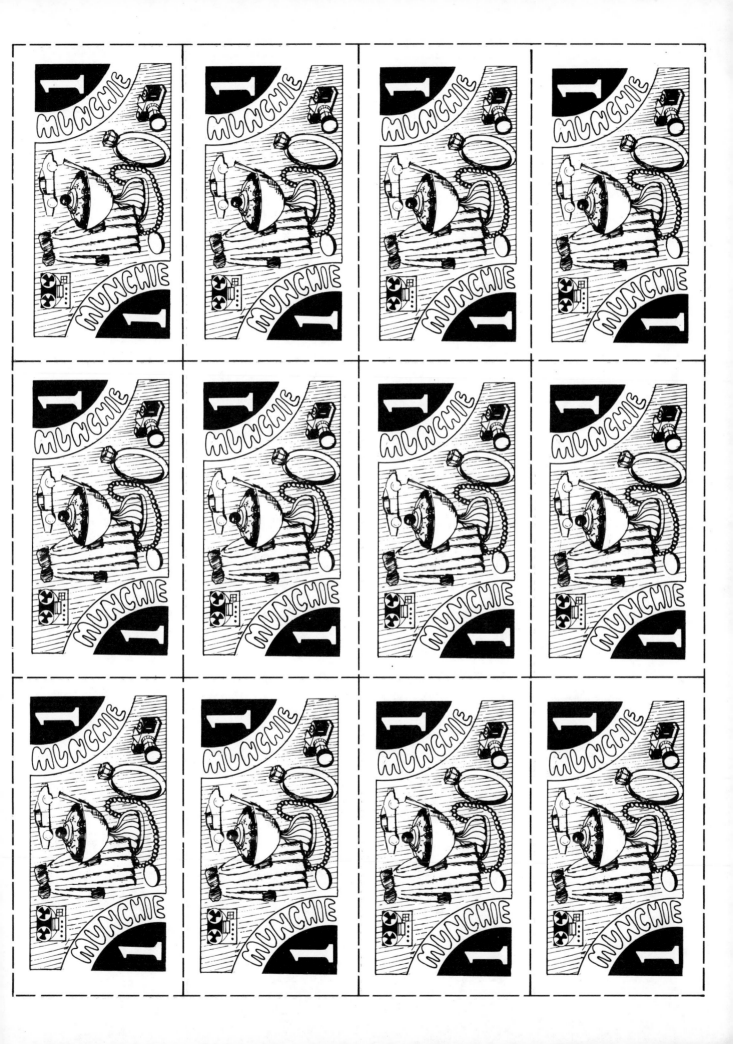

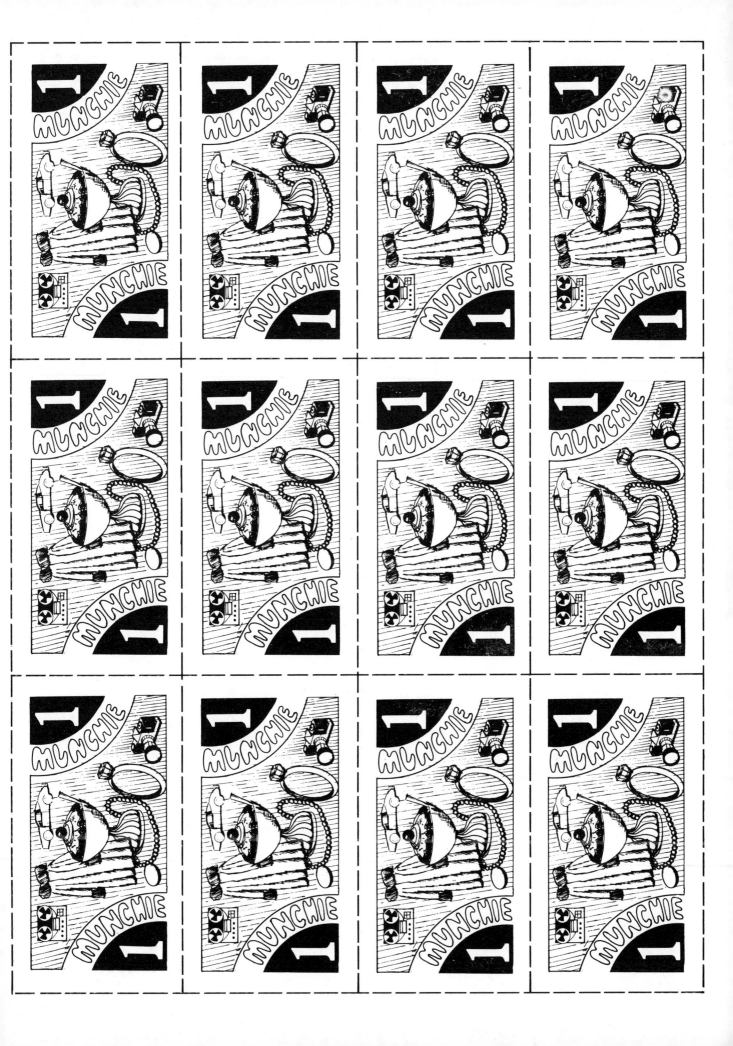

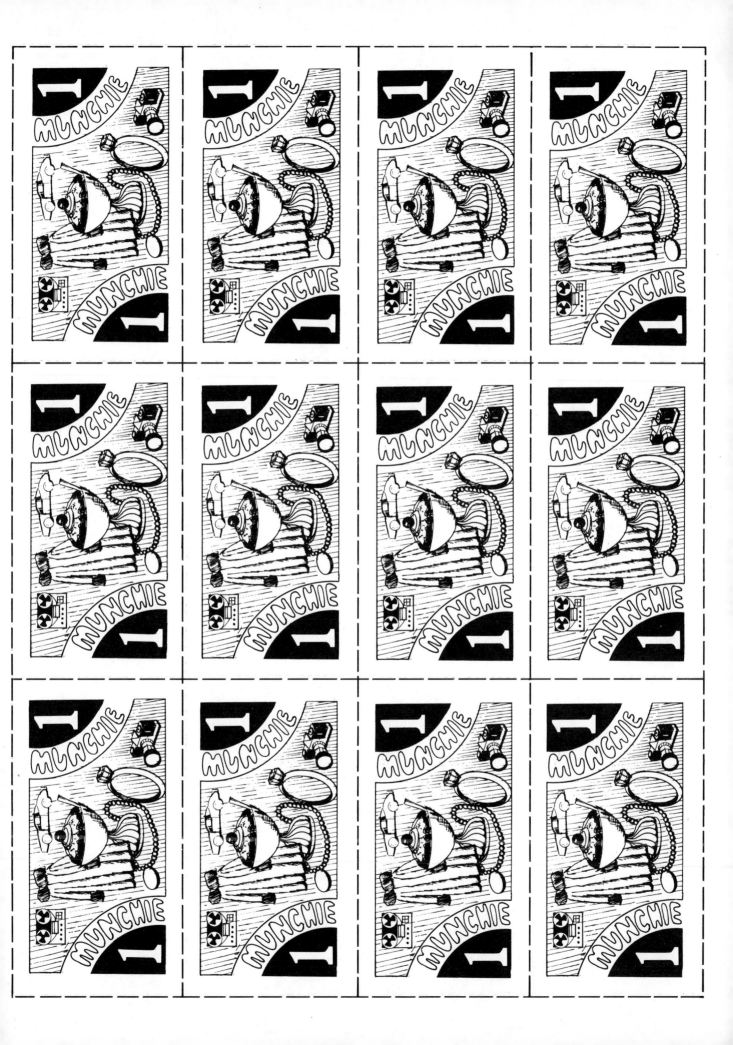

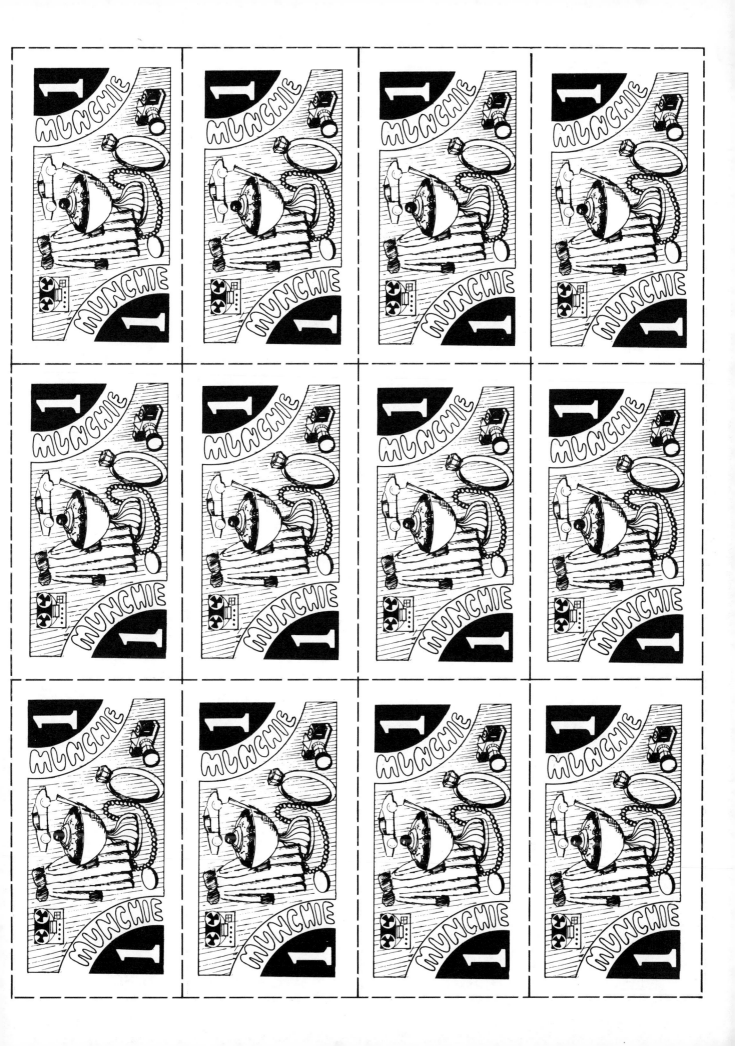

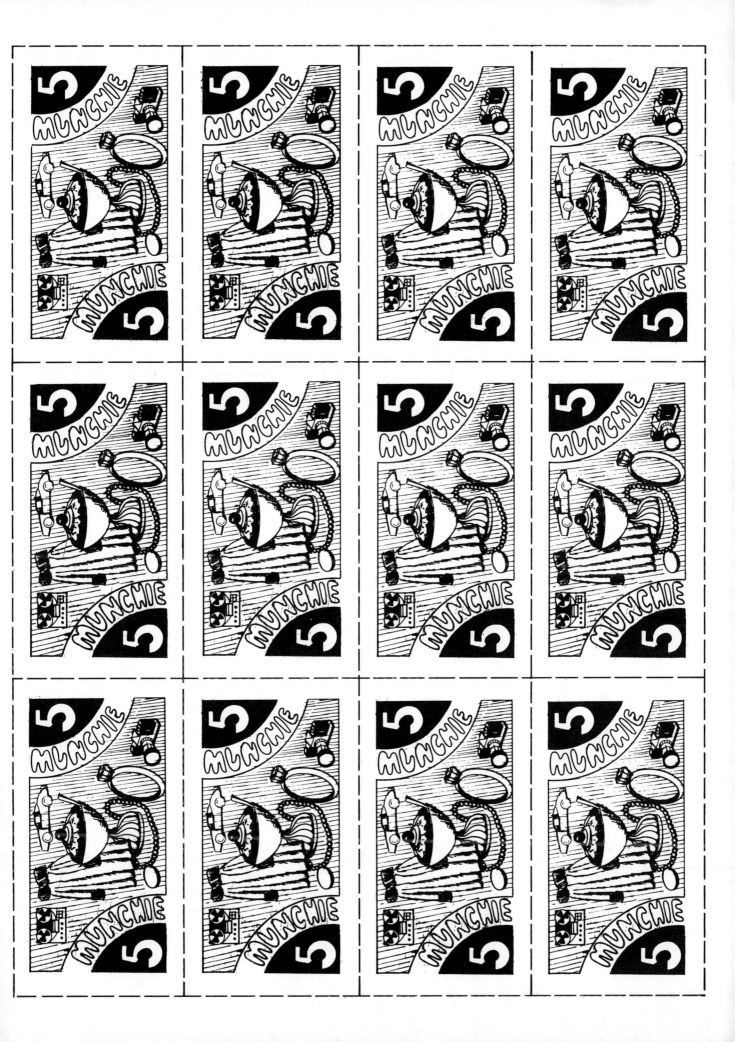

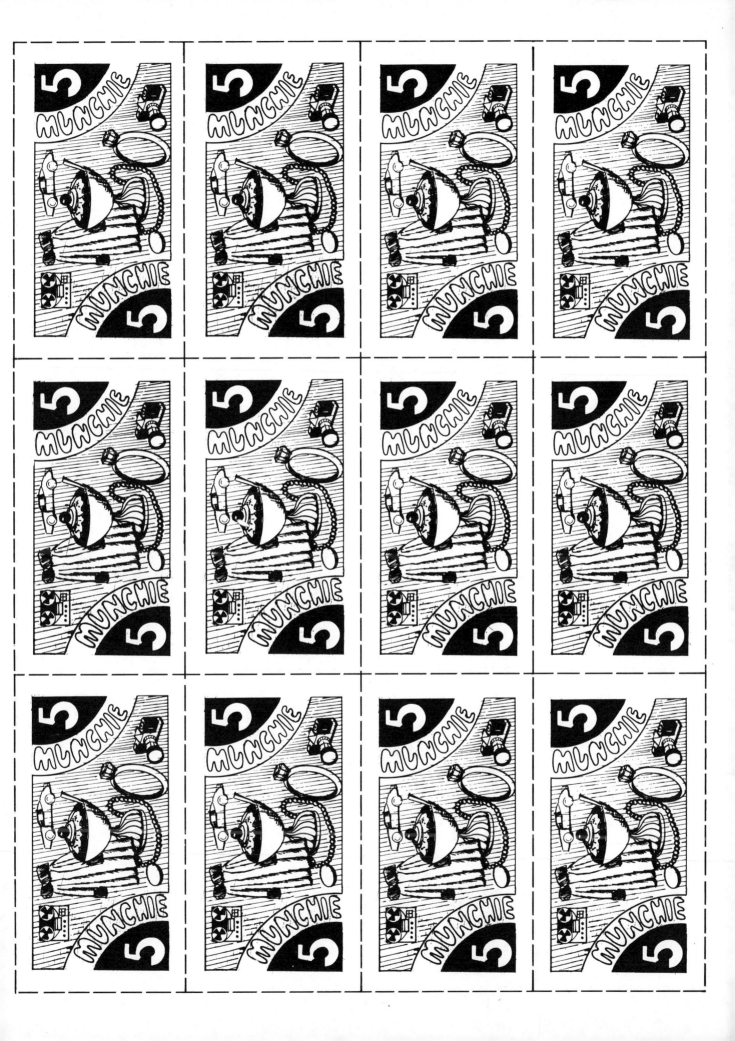

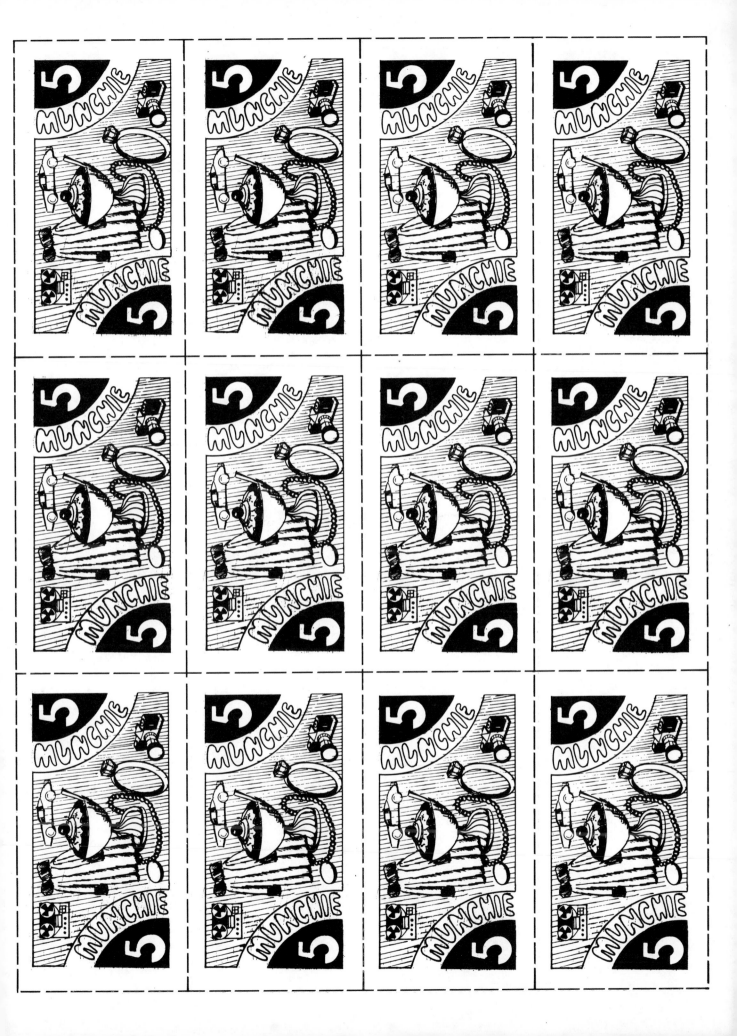

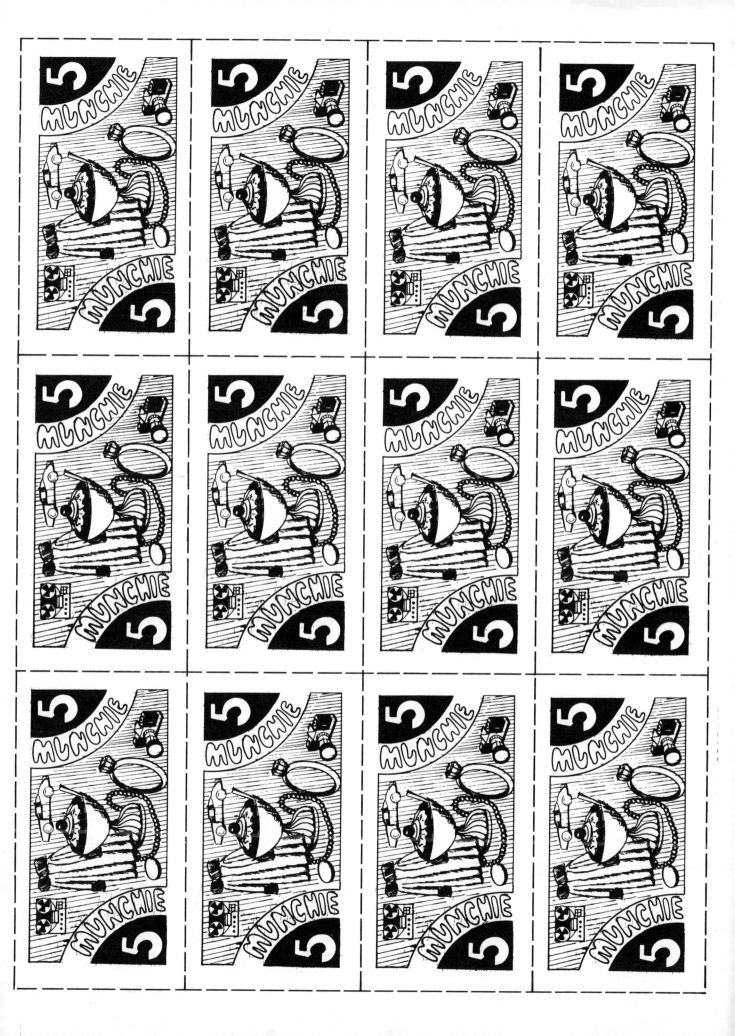